Machine Reliability and Condition Monitoring

A Comprehensive Guide to Predictive Maintenance Planning

Copyright © 2020 Mohammed Soliman

All rights reserved

Mohammed Hamed Ahmed Soliman

Contents

Introduction to Reliability and Reliability Centered Maintenance RCM 6

    Reliabiltiy Analysis 10

    Maintenance Policies and Strategies 12

    Predictive Maintenance Techniques 13

    Reliability KPIs 15

Equipment Criticality Analysis 16

Vibration Analysis 21

    Introduction and Terminologies 21

        On Site Tools Used For Measurements& Analysis of the Mechanical Vibration 33

        Measurement Techniques (points of measuring) 36

        Vibration Standards 39

        Quick Example – Centrifugal Fan 40

        Example.2 Discuss the Following Vibration Analysis Data Report 42

    Methodology of Measuring "Collecting Data" 46

    The Different Types of Vibration Sensors 47

        Vibration Sensors Connection Types 53

    Vibration Analysis-Signal Processing 63

        Time Waveform 64

- Enveloping and Demodulation Spectrum 66
- Phase Reading 67

Frequency Spectrum...67

What is FFT?..69

Spectrum Analysis & Faults Diagnosis 70

What are the Faults that Spectrum Analysis Can Tell Us About?...72

Faults to be Detected by Spectrum Analysis.........73

Predefined Spectrum Analysis Bands 74

Construction of Spectrum Analysis........................74

Spectrum Analysis Case Study: Detection of Different Failures in Rotating Machines...............77

- 1. Using Vibration Spectrum Analysis to Detect Machine Unbalance 79
- 2. Using Vibration Spectrum Analysis to Detect Bent Shafts..85
- 3. Using Vibration Spectrum Analysis to Detect Machine Misalignment 86
- Detect Transmission Belt & Sheaves Fault Condition Using Spectrum Analysis..................93

Case Study.2 "Detect electric Motor Faults by Vibration Spectrum Analysis" 99

- 1. Detect the Nature of Bearing Failure 102
- 2. Detection of Different Electric Problems 107

- Case study. 3 Using Vibration Spectrum Analysis to Detect Machine Looseness 116
- Case study.4: Detection of Bearing Looseness....120

Oil Analysis 123
- Oil Analysis Definition and Procedures 123
- Terminologies Involve in Lubricant Systems 126
- Role and Types of Different Oil Additives..........128
- Different Types of Oil Elemental Tests and Techniques Used 136
- Case Study.1: Analyzing Engine Oil 144
- Sampling Methods: 147
- Vibration VS Oil Analysis 150
- Case Study.2: Turbine Oil Condition Monitoring 152
- Case Study.3: Oil Condition Monitoring for Electrical Components 165

Thermography Analysis 169
- Introduction about Thermography Technique ..169
- Industrial Equipment Applications......................174
- Thermography Process Installations Application 183
- Electrical Systems Application 193
- Buildings Inspection 202

- Medical Applications 207
- Advantages of Thermography and Limitations . 213
- Infrared Thermometer – A simple hand held device 215
- How to Choose the Right Camera 219

Ultrasound Analysis 225
- Introduction to Ultrasound Technique 225
- Overview on the Instrument 227
- Typical Applications 228
- Advantages of Ultrasound 240
- Case Study: 244
- Ultrasonic Condition-Based Lubrication 249
- Faults Detection Using Ultrasound 259

References: 260

About the Author 262

Introduction to Reliability and Reliability Centered Maintenance RCM

The RCM philosophy employs Preventive Maintenance (PM), Predictive Testing and Inspection, Repair (also called reactive maintenance) and Proactive Maintenance techniques in an integrated manner to increase the probability that a machine or component will function in the required manner over its design life cycle with a minimum of maintenance. The goal of the philosophy is to provide the stated function of the facility, with the required reliability and availability at the lowest cost. RCM requires that maintenance decisions be based on maintenance requirements supported by sound technical and economic justification.

As with any philosophy, there are many paths, or processes, that lead to a final goal. This is especially true for RCM where the consequences of failure can vary dramatically. Rigorous RCM analysis has been used extensively by the aircraft, space, defense, and nuclear industries where functional failures have the potential to result in large losses of life, national security implications, and/or extreme environmental impact.

A rigorous RCM analysis is based on a detailed Failure Modes and Effects Analysis (FMEA) and includes probabilities of failure and system reliability calculations. The analysis is used to determine appropriate maintenance tasks to address each of the identified failure modes and their consequence. For more information about FMEA, read the book A Practical Guide to FMEA.

Equipment/Maintenance Reliability Definition

The instantaneous likelihoods of failure for a specific piece of equipment during a specific time period.

RCM Analysis

The RCM analysis carefully considers the following questions:
- What does the system or equipment do; what is its function?
- What functional failures are likely to occur?
- What are the likely consequences of these functional failures?
- What can be done to reduce the probability of the failure, identify the onset of failure, or reduce the consequences of the failure.

RCM Goals

- ✓ To ensure realization of the inherent safety and reliability levels of the equipment.
- ✓ To restore the equipment to these inherent levels when deterioration occurs.
- ✓ To obtain the information necessary for design improvement of those items where their inherent reliability proves to be inadequate.

- ✓ To accomplish these goals at a minimum total cost, including maintenance costs, support costs, and economic consequences of operational failures.

> Reliability centered maintenance (RCM) is a reliability tool that is used to ensure the inherent designed reliability of a process or piece of equipment through the understanding and discovery of equipment functions, functional failures, failure modes and failure effects. In performing a RCM analysis, the RCM team uses a structured decision process to develop mitigating tasks for each failure mode identified during the analysis.

Reliabiltiy Analysis

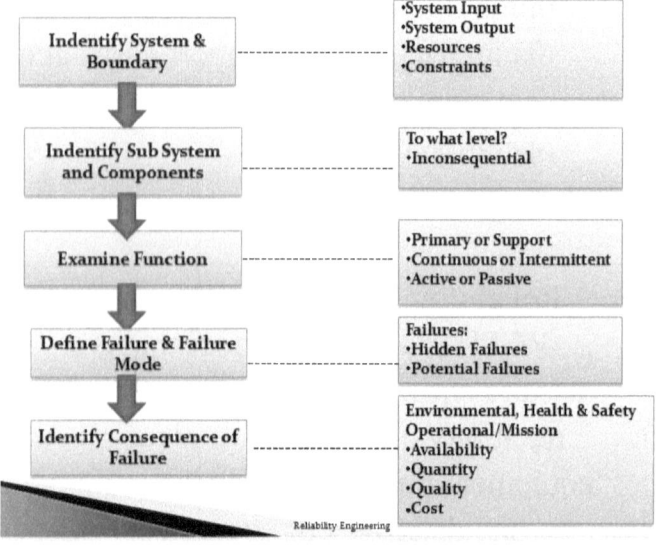

Maintenance Analysis

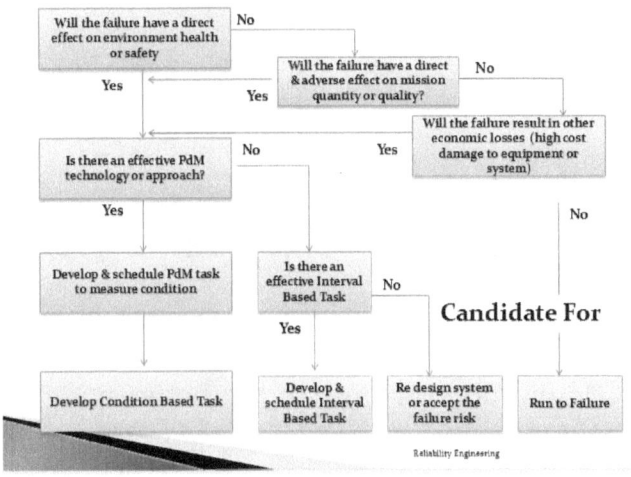

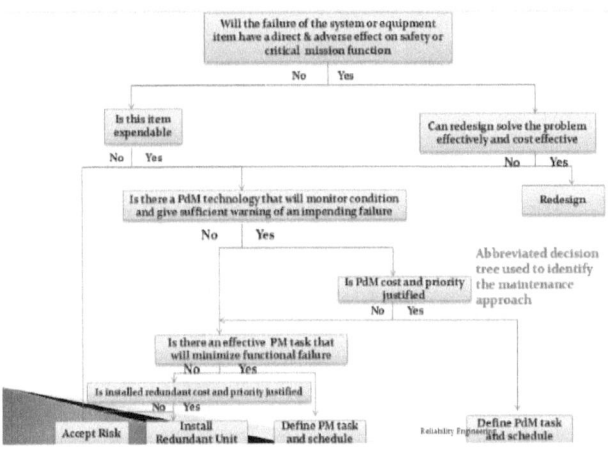

Abbreviated decision tree used to identify the maintenance approach

Maintenance Policies and Strategies

Maintenance Policies

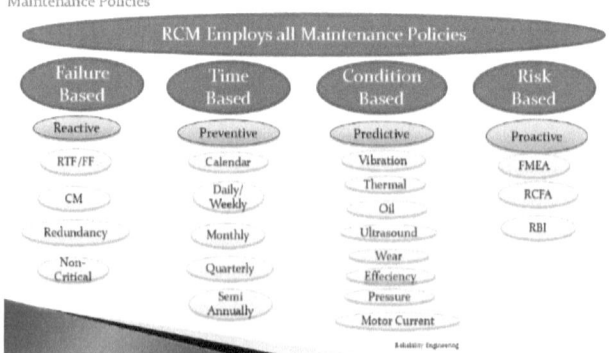

RCM promotes the use of Predictive and Risk Maintenance policies for identified critical equipment

Predictive Maintenance Techniques

Predictive Maintenance Embraced by Plant Maintenance

Technique	Application	Pumps	Electric Motors	Diesel Generators	Condensers	Heavy Equipment/ Crane	Circuit Breakers	Valves	Heat Exchangers	Electrical Systems	Transformers	Tank Piping
VIB Analysis		•	•	•		•						
Oil Analysis		•	•	•		•					•	
Wear Analysis		•	•	•		•						
IR Analysis		•	•	•	•	•	•	•	•	•	•	
Ultrasound		•	•	•			•	•	•	•	•	
Non-Destructive testing (Thickness)					•			•			•	
Visual Inspection		•	•	•	•	•	•	•	•	•	•	•
Motor Current Analysis		•										

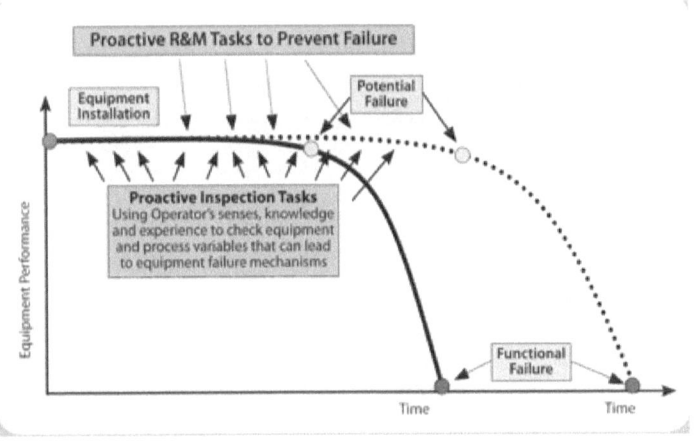

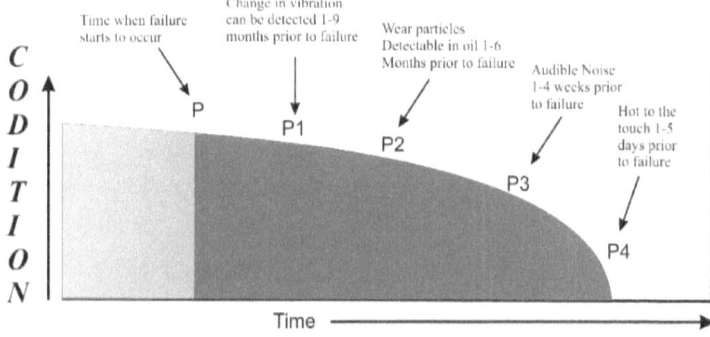

Reliability KPIs

KPI	Description
MTBF	Mean Time Between Failure
No of failures addressed by root cause analysis	>75%
Ratio of PM work orders to CM work orders generated by PdM inspection	
OEE (Overall Equipment Effectiveness)	Availability x Reliability x Quality (85%)
Percent of Faults Found in Predictive maintenance Survey (Vib, IR, UT, OA)	No of faults found/ No of devices checked (target <3%
Percent of equipment covered by condition monitoring	Target= 100%
Reliability of critical equipment	99%
Facility Availability	>98%
Availability of critical equipment	>98%
Percent emergency maintenance	<5%
Percent planned maintenance	90%

Equipment Criticality Analysis

Importance of Equipment Criticality Analysis:
1. Influence the priority assignment of the Work Orders.
2. Influence the Work Orders execution speed.
3. Determine which Maintenance Class should come first.
4. Effect the scheduling of the preventive maintenance program.
5. Influence the priority of the preventive maintenance work.
6. Help in determining which maintenance approach to be used.

Criticality Measuring Principles
1. **Safety**
- ✓ Safety equipment (equipment carrying peoples, firefighting system, and alike).
- ✓ Consequence of parts failure effect safety (elevator ropes broken, firefighting generator stopped…etc).

✓ Gas piping leakage at any point poses a risk.

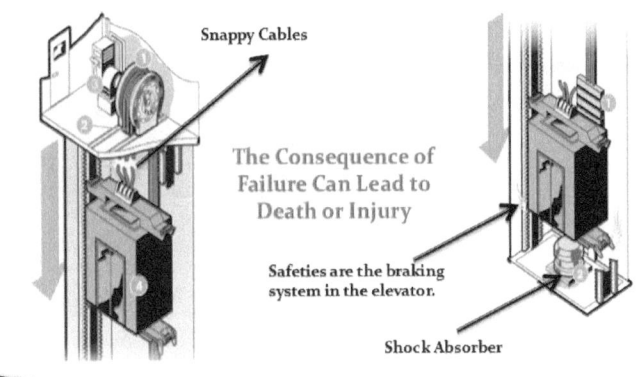

2. Production, Process
✓ Equipment breakdown affect the whole production line.
✓ Equipment breakdown affect partially the production line.
✓ Equipment failure has native effect on production quality.

The Consequence of Failure Can Lead to Process Shutdown, Major Losses, or Quality Problem.

Criticality decreases with redundancy in the system.

Criticality is influenced by the availability of standby equipment in a system but how much time does it take for the standby equipment to operate? And what is the effect of this on the production?

3. Time (Time=Money)
1 min production=How much?
1 Hr production =How much?

If your equipment is classified as critical, ask yourself the following questions:

- What is the preventive maintenance program I have for it? Enough? Or not?
- How much time does it take to repair it in case of failure?
- Spare parts allocation? Available? Not available? If not available, how much does it take to allocated it from the vendor?
- Require special skills for repair? Is my team trained to repair it in a proper time?
- Do you have an emergency plan for it in case of accident or failure? Is your team aware of this plan and trained on it?

4. Money, Cost

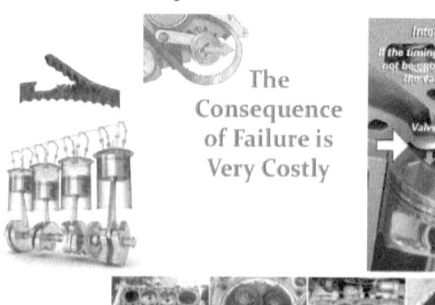

The Consequence of Failure is Very Costly

Equipment that fails in service can cost up to 10 times more repair than equipment repaired wh predicted b condition monitoring

Vibration Analysis

Introduction and Terminologies
What is Vibration?
Vibration can be defined as simply the cyclic or oscillating motion of a machine or machine component from its position of rest.

Displacement:
It is the amplitude of a point on a structure.

Velocity:
Is the speed of a point in a system, It is the rate of change of displacement.

Acceleration:
Is the rate of change of velocity of a point in a system.

Frequency:
Is the no of cycles (vibrations) per second, measured by hertz (HZ).

Harmonics:
Frequency component at a frequency that is an integer (whole number e.g. 2X. 3X. 4X,

etc) multiple of the fundamental (reference) frequency.

How many RPMs in 1 hertz?
Since hertz is in sec, and RPM is in minute, 1 hz= 60 RPM.

How to convert RPM to Hertz?
1500RPM = 25hz

Why we measure the vibration?
- ✓ To detect what is out of the human sense.
- ✓ To discover hidden failures.
- ✓ To Detect early failures & monitor the machine
 health condition.
- ✓ To assure the quality of repairs.
- ✓ As a useful tool to improve the maintenance
 Reliability.

Common Industrial Applications:
- Pumps
- Fans
- Turbines
- Agitators
- Stirrers
- Compressors
- Electric Motors
- Gearboxes

Vibration& Reliability

Vibration analysis alone doesn't improve reliability, root cause analysis and acceptance testing can help.

There are two ways that we can utilize vibration to improve reliability:

First, if we study the vibration we can often determine why the fault condition developed in the first place; for example what caused the crack to appear in the inner race of the bearing? If we perform root cause failure analysis we can make chances to our proactive so that the bearing don't suffer the same fate in the future.

Second, when we overhaul the machine, we can again use the vibration analysis to check that the maintenance repair has been made correctly; and that the machine is correctly aligned and balanced, this called acceptance testing.

Vibration is still used to monitor the health of the machine, but if we improved the reliability of the machine we will see fewer faults conditions develop.

Measuring With Smart Sensors
"Collecting Data"

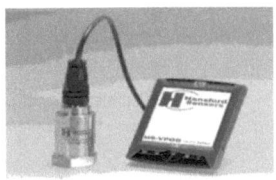

Analysis with Smart Software

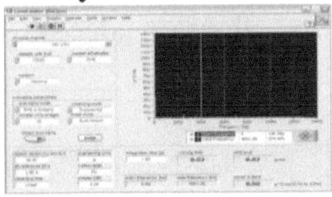
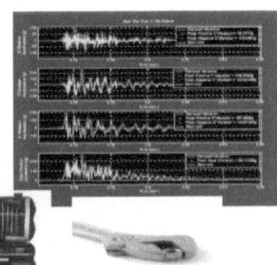

Getting Results

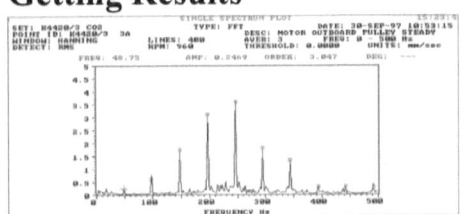

Performing Actions accordingly
Maintain

Repair
Cure

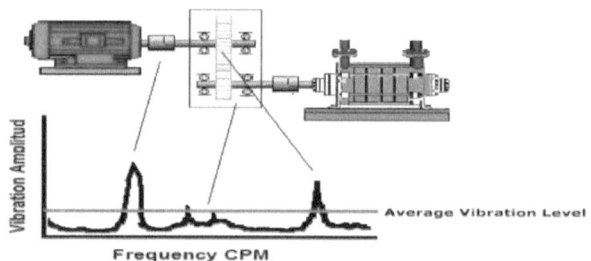

What to Measure??

We normally measure the vibration speeds in in/sec or mm/s.

Why using Condition Monitoring programs (predictive maintenance)? And why particularly Vibration Analysis technique?

Benefits of setting up a Predictive Maintenance (PdM) program:
1. To detect what is out of the human sense.
2. To discover hidden failures.
3. To Detect early failures & monitor the machine health condition.
4. To reduce Maintenance Costs.
5. As a useful tool to improve the machine reliability.

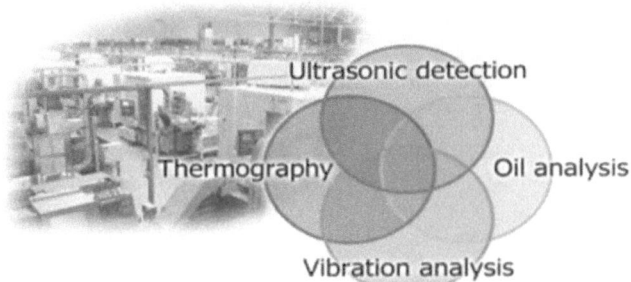

Four tools make up 85% of any PdM program

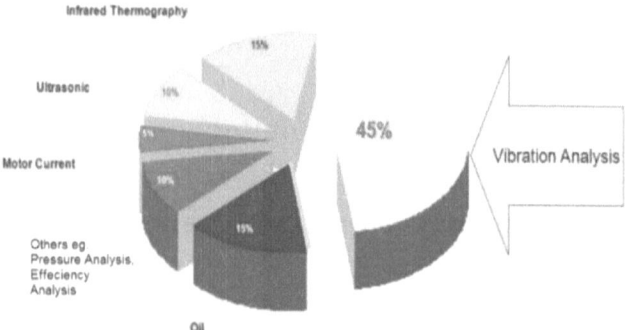

Vibration present 45% of PdM programs

Equipment that fails in service can cost up to 10 times more to repair than the equipment repaired when predicted by condition monitoring.

Other PdM Techniques

Notes:
Motor diagnosis = motor current analysis, and it's a technique involve intensive diagnosis of motor currents.

Oil Analysis involve Wear Particles Analysis for more intensive diagnosis about the sources of failure. For more information about the technique read the book: Machinery Oil Analysis and Condition Monitoring.

Thermography: involve thermal analysis using infrared camera. For more information about the technique, read the book: Industrial Applications of Infrared Thermography.

Ultrasound Analysis: is an acoustic method based on high frequencies measurement. For more information, read the book: Ultrasound Analysis for Condition Monitoring.

Why Vibration?

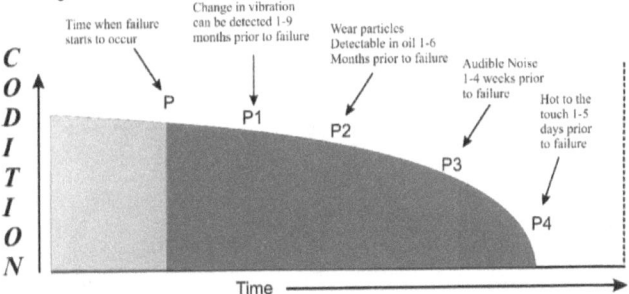

Vibration VS Thermography VS Oil Analysis

Type of fault	Vibration	Temp	Oil
Out of balance	xxx	----	----
Misalignment	xxx	x	----
Damage of bearing	xxx	xx	x
Damage of gear box	xxx	x	xx
Belt problems	xx	----	----
Motor problems	xx	x	----
Mechanical looseness	xxx	x	x
Resonance	xxx	----	----

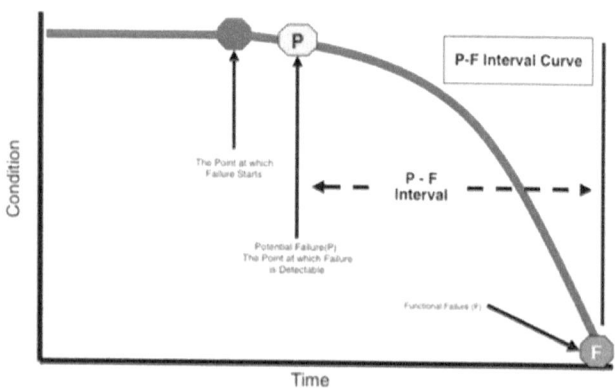

One of the most benefits of a condition monitoring program is to detect potential failures at early state

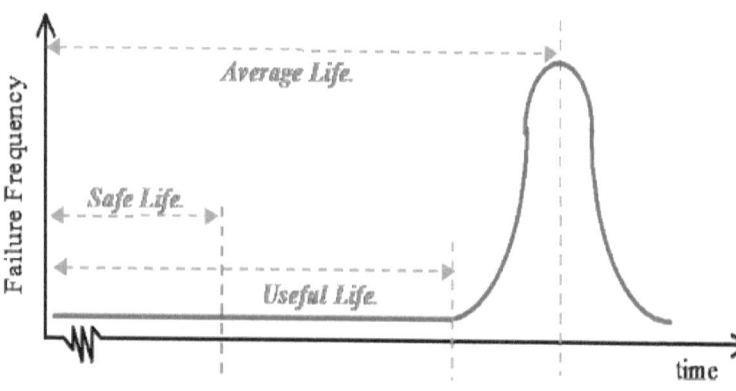

Determine the PM Interval Using Reliability Data from PdM Programs

On Site Tools Used For Measurements & Analysis of the Mechanical Vibration

1. **Simple tools, Vibration pen**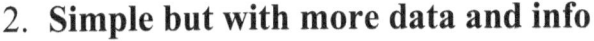
   Simple.
   Accurate.
   Easy.
   Less Expensive.
   Quick Measure.

2. **Simple but with more data and info**
   ✓ Quick spectrum view.
   ✓ Compare vibration reading to the charts that indicate the severity of the faults according to the type and size of the machine.
   ✓ Inexpensive way to start with vibration measurement.

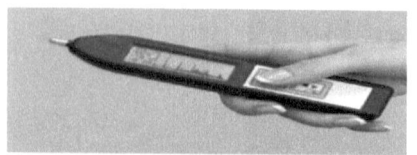

Machines Vibration Limits and Reference

Class I (Small): Machines less than 20HP.
Class II (Medium): Machines from 20-100HP without special foundations.
Class III (Large): Machines with rigid foundations and over 100HP.
Class IV (Large): Machines with soft foundation and over 100HP.

VIBRATION SEVERITY PER ISO 10816					
Machine		Class I small machines	Class II medium machines	Class III large rigid foundation	Class IV large soft foundation
in/s	mm/s				
0.01	0.28				
0.02	0.45				
0.03	0.71	good			
0.04	1.12				
0.07	1.80				
0.11	2.80	satisfactory			
0.18	4.50				
0.28	7.10	unsatisfactory			
0.44	11.2				
0.70	18.0				
0.71	28.0	unacceptable			
1.10	45.0				

3. Professional Detectors/Analyzers

- ✓ Professional.
- ✓ More Accurate.
- ✓ Software Analysis.
- ✓ Automatic Analysis.
- ✓ Professional results.
- ✓ Costly.

Please discuss with the vendor the suitable instrument for your business!

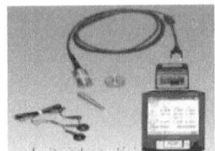 

4. **Portable** professional analyzers with ERP connected:
   - ✓ Dual Channel Analyzer.
   - ✓ Single Channel Analyzer.
   - ✓ USB Connectivity.

Measurement Techniques (points of measuring)

Measuring at bearing points in three directions:
Horizontal
Vertical
Axial

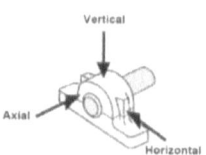

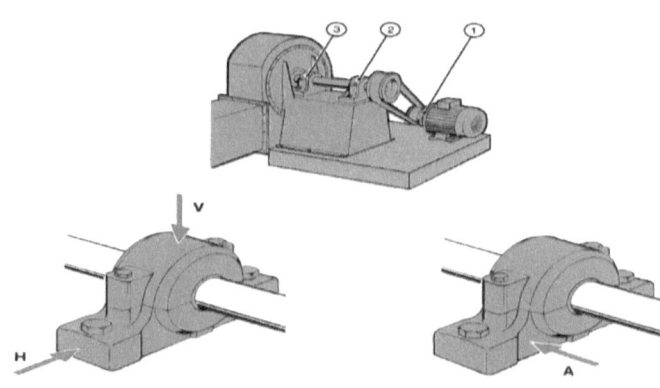

Why measure in 3 directions?
Taking readings in three directions gives more information that helps in analysis

as some defects comes with predominant vibrations in a particular directions.

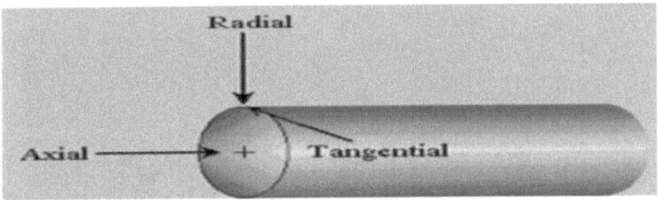

Axial is the direction parallel to the centerline of a shaft or turning axis of a rotating part. Radial is that direction toward the center of rotation of a shaft or rotor. The Tangential measurement is that measurement that is tangent or perpendicular to the radial transducer.

Sensor positions, what is right and what is wrong

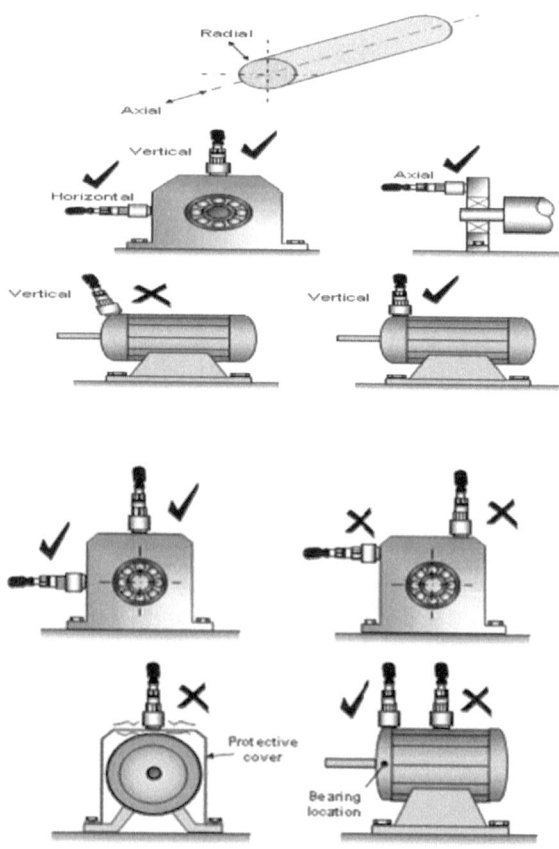

Vibration Standards

Vibration Reports & Standards

Standard acc to ISO 10816

VIBRATION SEVERITY PER ISO 10816					
Machine		Class I small machines	Class II medium machines	Class III large rigid foundation	Class IV large soft foundation
in/s	mm/s				
0.01	0.28				
0.02	0.45				
0.03	0.71	good			
0.04	1.12				
0.07	1.80				
0.11	2.80	satisfactory			
0.18	4.50				
0.28	7.10	unsatisfactory			
0.44	11.2				
0.70	18.0	unacceptable			
0.71	28.0				
1.10	45.0				

Class I (Small): Machines less than 20HP.

Class II (Medium): Machines from 20-100HP without special foundations.

Class III (Large): Machines with rigid foundations and over 100HP.

Class IV (Large): Machines with soft foundation and over 100HP.

Quick Example – Centrifugal Fan

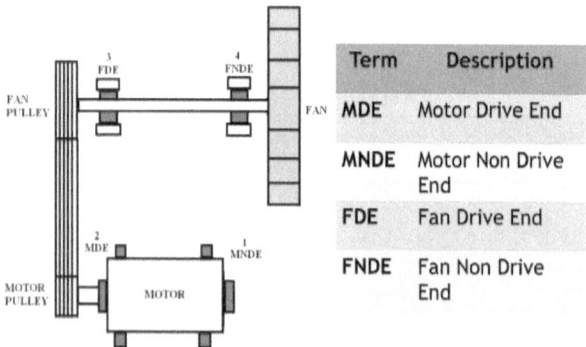

Term	Description
MDE	Motor Drive End
MNDE	Motor Non Drive End
FDE	Fan Drive End
FNDE	Fan Non Drive End

Ex. General Components of centrifugal fan

Vibration Measurements

POINT	DESCRIPTION	overall values	LIMITS
1 H	E-Motor non drive end horizontal	30.468 mm/s	x x x
1 V	E-Motor non drive end vertical	13.611 mm/s	x x x
1 A	E-Motor non drive end axial	-	
2 H	E-Motor drive end horizontal	24.302 mm/s	x x x
2 V	E-Motor drive end vertical	23.217 mm/s	x x x
2 A	E-Motor drive end axial	22.827 mm/s	x x x
3 H	fan fixed bearing coupling side horizontal	35.610 mm/s	x x x
3 V	fan fixed bearing coupling side vertical	31.521 mm/s	x x x
3 A	fan fixed bearing coupling side axial	-	
4 H	Fan free bearing fan side horizontal	29.609 mm/s	x x x
4 V	Fan free bearing fan side vertical	26.941 mm/s	x x x
4 A	Fan free bearing fan side axial	24.733 mm/s	x x x

DANGEROUS	x x x
ALARM	x x
ACCEPTED	x

Spectrum chart

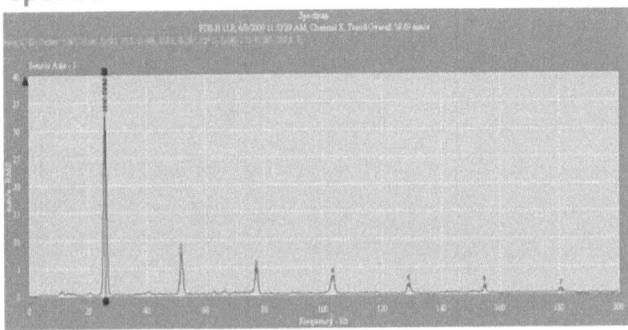

Fig.

Machine showing significant increase in vibration level at frequency of 25hz indicating unbalance issue.

Action report/recommendations examples:

Perform Balancing.

Need to Replace Bearing no 1, 2...etc.

Check for Bearing Grease & lubrication condition.

Monitor Bearing Condition for Early Failure.

Perform Alignment (for shaft, pulleys…etc.).

Check for Foundation Looseness, internal looseness…etc.

Example.2 Discuss the Following Vibration Analysis Data Report

Equipment: Centrifugal Fan
Class: IV
Measurements are in mm/s

Parameters		1-9-2010	1-10-2010	1-11-2010	1-12-2010	1-1-2011
Point 1 (fan drive end)	x	1.411	2.835	4.7	9.5	12.7
	y	1.865	2.80	4.2	8.0	14.3
	z	2.487	4.853	4.9	7.3	7.2
Point 2 (fan non drive end)	x	1.490	3.0	4.87	6.5	14.0
	y	0.9	1.0	2.5	4.2	7.0
	z	1.854	2.2	4.1	7.1	12.0

System Description
Technical Data:
Fan Type= Centrifugal fan
Flow rate= 78,000m3/h
Motor Power= 200KW
Horse power= 268HP
Motor Speed= 1500RPM
Fan Speed= 1500RPM

Transmission type: V-belts

Spare Parts Data:

FDE Bearing: 22222EK +H322

FNDE Bearing: 22222EK +H322

MDE Bearing: 6322

MNDE Bearing: 6322

Bearing Housing: SNH 522-619

Pulleys type: SPC335

V-belts type: SPC5300

Equipment: Centrifugal Fan
Class: IV
Measurements are in mm/s

Parameters		1-9-2010	1-10-2010	1-11-2010	1-12-2010	1-1-2011
Point 1 (fan drive end)	x	1.411	2.835	4.7	9.5	12.7
	y	1.865	2.80	4.2	8.0	14.3
	z	2.487	4.853	4.9	7.3	7.2
Point 2 (fan non drive end)	x	1.490	3.0	4.87	6.5	14.0
	y	0.9	1.0	2.5	4.2	7.0
	z	1.854	2.2	4.1	7.1	12.0

Green color = Acceptable
Yellow color = Alarm
Red color = Dangerous

Remarks:
First month: Over all machine health is good.
Second month: Over all machine health is within the acceptable limits "but recommend to monitor bearing condition"
Third month: There is a significant increase in vibration speed, machine will require balancing.
Forth month: machine is in critical level and required immediate shutdown to perform the following:
 -Balancing.
 -checking for pulleys alignment and belt tensioning.
 -Monitor non-drive end bearing and drive end bearing for replacing .requirement or greasing.
Fifth month: Stop the machine to avoid sudden failure caused by one of the following issues:-
 -Crack in foundation.
 -Bearing failure.
 -Housing wear.

Question: Why it's important to have all the machine data & spare parts information before processing the vibration analysis?

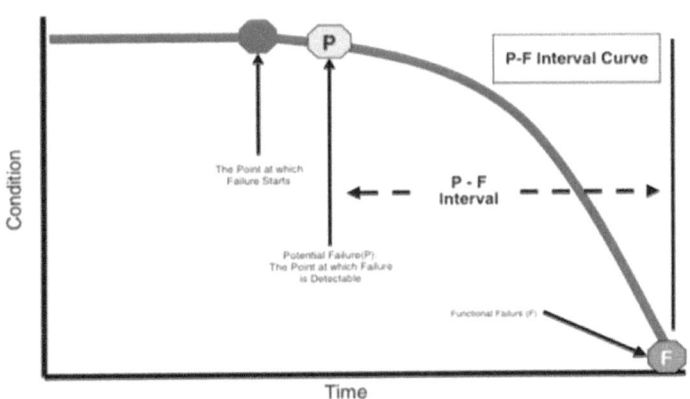

One of the most benefits of a Condition Monitoring program is to detect potential failures at early state.

Methodology of Measuring "Collecting Data"

By using a sensor called accelerometer & an electronic meter that record the vibration.

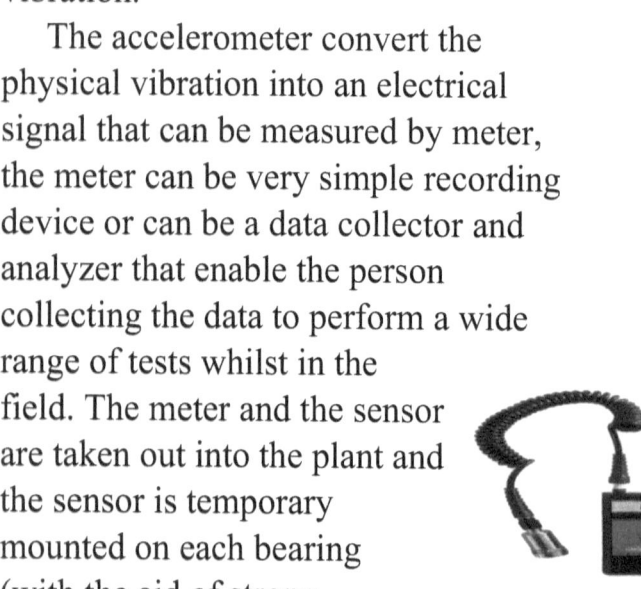

The accelerometer convert the physical vibration into an electrical signal that can be measured by meter, the meter can be very simple recording device or can be a data collector and analyzer that enable the person collecting the data to perform a wide range of tests whilst in the field. The meter and the sensor are taken out into the plant and the sensor is temporary mounted on each bearing (with the aid of strong magnet) and the vibration reading is taken.

A Simple meter display the vibration reading for comparison against alarms, but more advanced meters record the vibration for later analysis.

The Different Types of Vibration Sensors
1-Velocity pickup
The velocity pickup is a very common transducer for monitoring the vibration of rotating machinery. This type of vibration transducer installs easily on most analyzers, and is rather inexpensive compared to other sensors. For these reasons, the velocity transducer is ideal for machine-monitoring. Velocity pickups have been used as vibration transducers on rotating machines.

For a very long time, and these are still utilized for a variety of applications today. Velocity pickups are available in many different physical configurations and output sensitivities.

Theory of Operation
When a coil of wire is moved through a magnetic field (coil-in-magnet type) a voltage is induced across the end wires of the coil. The transfer of energy from the flux field of the magnet to the wire coil generates the induced voltage. As the coil is forced

through the magnetic field by vibratory motion, a voltage signal correlating with the vibration is produced.

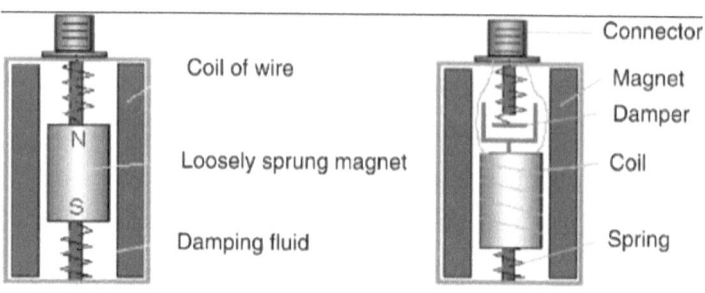

The magnet-in-coil type of sensor is made up of three components: a permanent magnet, a coil of wire and spring supports for the magnet. The pickup is filled with oil to dampen the spring action. The relative motion between the magnet and coil caused by the vibration motion induces a voltage signal.

2-Acceleration transducers/pickup

Accelerometers are the most popular transducers used for

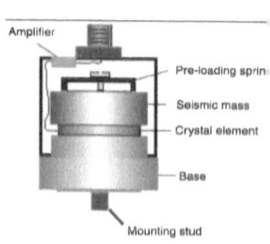

rotating machinery applications. They are rugged, compact, lightweight transducers with a wide frequency response range. Accelerometers are extensively used in many condition-monitoring applications. Components such as rolling element bearings or gear sets generate high vibration frequencies when defective. Machines with these components should be monitored with accelerometers.

The installation of an accelerometer must carefully be considered for an accurate and reliable measurement.

ACCELEROMETER: Transducer whose output is directly proportional to acceleration. Most commonly used are mass loaded piezoelectric crystals to produce an output proportional to acceleration.

Theory of operation

Accelerometers are inertial measurement devices that convert mechanical motion into a voltage signal. The signal is proportional to the vibration's acceleration using the piezoelectric principle. Inertial measurement devices measure motion relative to a mass. This follows Newton's third law of motion: body acting on another will result in an equal and opposite reaction on the first. Accelerometers consist of a piezoelectric crystal (made of ferroelectric materials like lead zirconate titanate and barium titanate) and a small mass normally enclosed in a protective metal case.

When the accelerometer is subjected to vibration, the mass exerts a varying force on the piezoelectric crystal, which is directly proportional to the vibratory acceleration. The charge produced by the piezoelectric crystal is proportional to the varying vibratory force. The charge output is measured in Pico-coulombs per g (Pc/g) where g is the gravitational acceleration. Some sensors have an internal charge amplifier, while others have an external charge amplifier. The charge amplifier converts the charged output of the crystal to a proportional voltage output in mV/g.

Crystals will generate measurable piezoelectricity when their static structure is deformed by about 0.1% of the original dimension.

Frequency range

Accelerometers are designed to measure vibration over a given frequency range. Once the particular frequency range of interest for a machine is known, an accelerometer can be ranges are also available selected. Typically, an accelerometer for measuring machine vibrations will have a frequency range from 1 or 2 Hz to 8 or 10 kHz. Accelerometers with higher-frequency

Calibration for Both Types

Piezoelectric accelerometers cannot be recalibrated or adjusted. Unlike a velocity pickup, this transducer has no moving parts subject to fatigue. Therefore, the output sensitivity does not require periodic adjustments. However, high temperatures and shock can damage the internal components of an accelerometer.

Velocity pickups should be calibrated on an annual basis. The sensor should be removed from service for calibration verification. Verification is necessary because velocity pickups are the only industrial vibration sensors with internal moving parts that are subject to fatigue failure.

Vibration Sensors Connection Types
Wireless

Wired

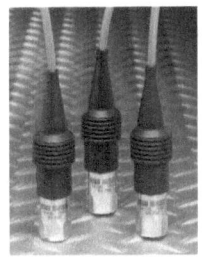

Wired Vibration Sensors Limitation:
- ✓ High cost of installation, especially in hazardous areas.
- ✓ Insufficient justification for a permanent system on certain balance-of-plant machines.
- ✓ Traversing/moving machines where fixed cabling is not possible.

Wireless Vibration Sensors Advantages:
- ✓ Overcome the wire sensor limitation.
- ✓ Facilitate applications that in the past were impractical, such as temporary

installations for troubleshooting and remote monitoring.
- ✓ Eliminates the need for hardwiring for communications.
- ✓ Reduced installation cost.

Challenges to Wireless Vibration Sensors:
- ✓ High bandwidth is needed, due to the relatively large amounts of data that need to be sent over the wireless link.
- ✓ Higher-level processing capabilities, and the ability to capture data at the right time are also key requirements.
- ✓ Battery-powered devices that are required to provide onboard power must satisfy customer demands for long service life.
- ✓ Wireless security is a must.

Tips:
The devices and sensors, as well as the wireless network components, must also cope with conditions commonly found in the industrial environment, such as exposure to

water, elevated temperatures, electrical interference, hazardous-area classifications, obstructions, physical location and distance.

 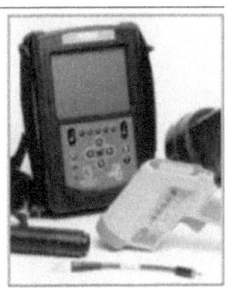

Advantages
- ✓ Can collect, record and display vibration data such as FFT spectra, overall trend plots and time domain waveforms.
- ✓ Provides orderly collection of data.
- ✓ Automatically reports measurements out of pre-established limit thresholds.
- ✓ Can perform field vibration analysis.

Disadvantages
- ✓ They are relatively expensive.
- ✓ Operator must be trained for use.

- ✓ Limited memory capability and thus data must be downloaded after collection.

Vibration analysis – database management software

The data collector/analyzer can collect and store only a limited amount of data. Therefore, the data must be downloaded to the computer to form a history and long-term machinery information database for comparison and trending. To perform the above tasks, and also aid in collection, management and analysis of machinery data, database management software packages are required.

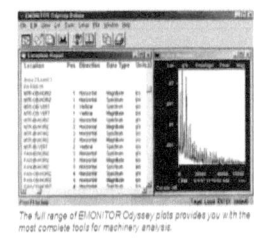

The full range of EMONITOR Odyssey plots provides you with the most complete tools for machinery analysis.

These database management programs for machinery maintenance store vibration data and make comparisons between current measurements, past measurements and predefined alarm limits. Measurements transferred to the vibration analysis software are rapidly investigated for deviations from normal conditions. Overall vibration levels, FFTs, time waveforms.

Reports can be generated showing machines whose vibration levels cross alarm thresholds. Current data are compared to baseline data for analysis and also trended to show vibration changes over a period of time. Trend plots give early warnings of possible defects and are used to schedule the best time to repair.

Advantages
- They aid in data collection, management and analysis of machinery data.
- They can save long-term machinery data that help to compare present and past condition-monitoring data.
- They assist in vibration analysis.
- They provide user-friendly reports.

Disadvantages
- The software programs are expensive, with sometimes almost the same cost as the data collector/analyzer hardware.

- They must be configured for individual requirements. A lot of information is required as initial input.

Online data acquisition and analysis (for Critical Machines)

Critical machines, as defined in the earlier topics, are almost always provided with continuous online monitoring systems. Here sensors (e.g. Eddy current probes installed in turbo-machinery) are permanently installed on the machines at suitable measurement positions and connected to the online data acquisition and analysis equipment. The vibration data are taken automatically for each position and the analysis can be displayed on local monitoring equipment, or can be transferred to a host computer installed with database management software.

This ability provides early detection of faults and supplies protective action on critical machinery.

Advantages
- ✓ Performs continuous, online monitoring of critical machinery.
- ✓ Measurements are taken automatically without human interference.
- ✓ Provides almost instantaneous detection of defects.

Disadvantages
- ✓ Reliability of online systems must be at the same level as the machines they monitor.
- ✓ Failure can prove to be very expensive.
- ✓ Installation and analysis require special skills.
- ✓ These are expensive systems.

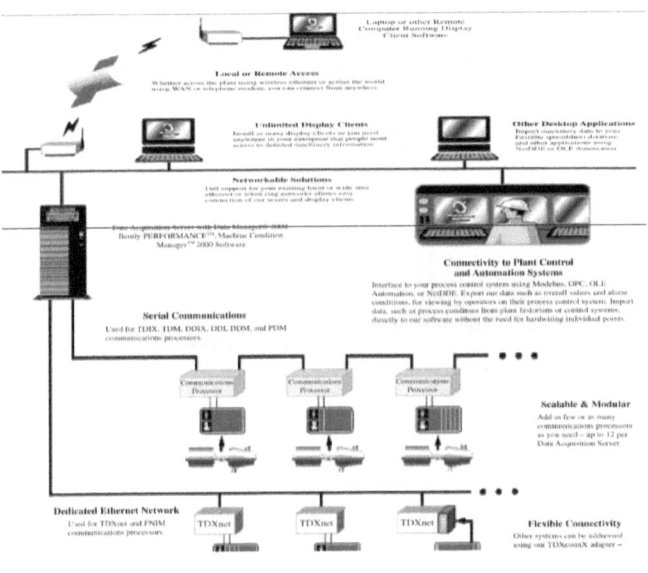
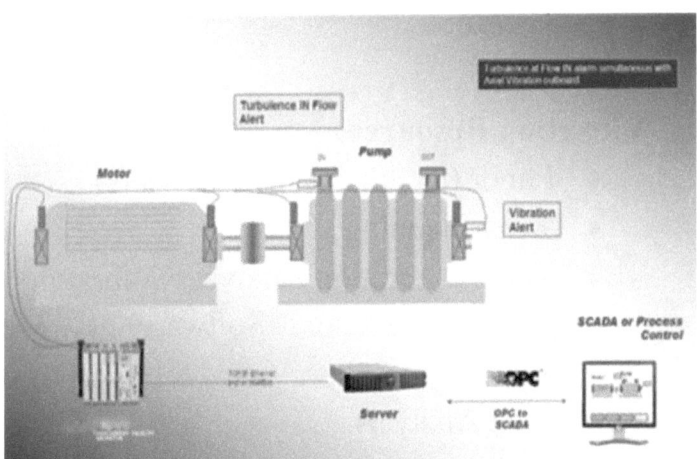

Vibration can be made online so you can have permanent monitoring to the critical machines health & condition.

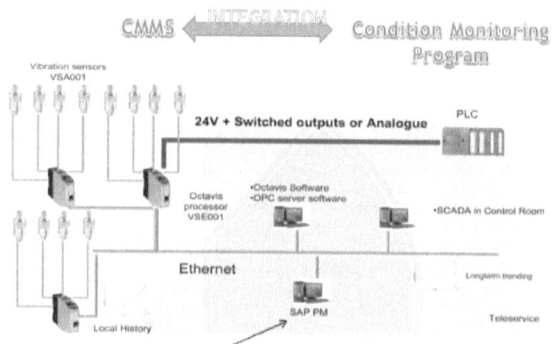

Automatic initiation of work orders depend on the machine condition

Vibration Resources
- Labor (technicians & engineers).
- Time Planning.
- Tools (Vibration Analyzer/ Vibration detector + Sensor + Software).
- Training.
- Cost Analysis.
- Budget planning & preparation.

Vibration Analysis-Signal Processing

The vibration of a machine is a physical motion. Vibration transducers convert this motion into an electrical signal. The electrical signal is then passed on to data collectors or analyzers. The analyzers then process this signal to give the FFTs and other parameters.

To achieve the final relevant output, the signal is processed with the following steps:

- Analog signal input
- Anti-alias filter
- A/D converter
- Overlap
- Windows
- FFT
- Averaging
- Display/storage.

Time Waveform

A time waveform is the time domain signal. In vibration terms, it is a graph of displacement, velocity or acceleration with respect to time. The time span of such a signal is normally in the millisecond range. A graphical representation of a wave obtained by plotting some characteristic of the wave (such as amplitude), versus time.

The raw signal from the sensor is the time waveform

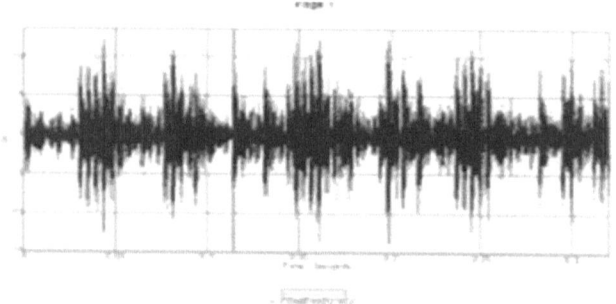

It's quite difficult to understand the time waveform
Time Waveform is an important analysis tool in certain circumstances.

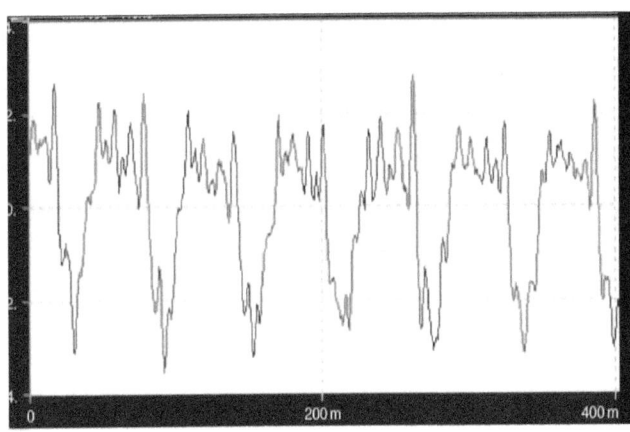

Enveloping and Demodulation Spectrum
This technique of vibration analysis is extensively used for fault detection in bearings and gearboxes. This method focuses on the high-frequency zone of the spectrum. Using a high-pass filter (allows high frequencies but blocks lower ones), the analyzer zooms into the low-level high-frequency data.

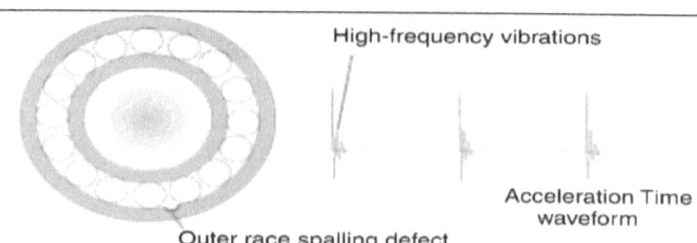

Phase Reading

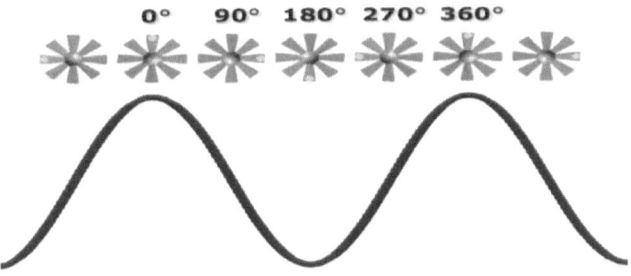

Frequency Spectrum

A spectrum is a graphical display of the frequencies at which a machine component is vibrating.

It's the FFT of the time waveform which produce the spectrum.

The data collector take the raw time waveform from the sensor and perform a calculation called FFT (Fast Fourier Transform). This process extract all of the individual frequencies from the vibration pattern.

When machines run, it generate vibrations at different frequencies, the shaft turn is one frequency, the vans on the pump generate higher frequency, and the bearing generate another frequency.

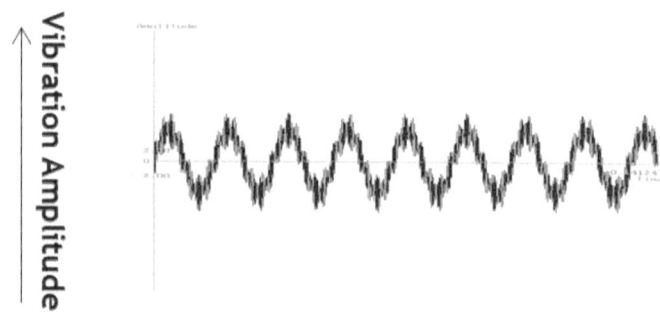

Signal Processing Flow

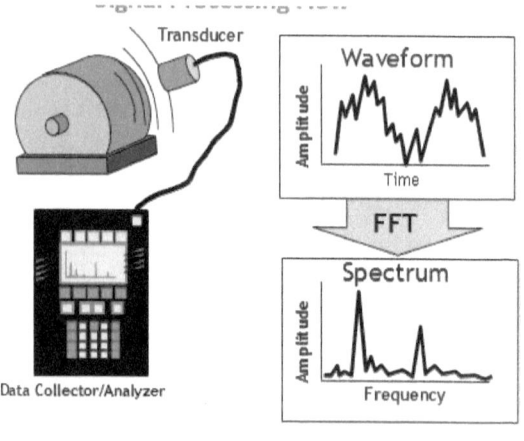

What is FFT?

The Fast Fourier Transform is a mathematical method for transforming a function of time into a function of frequency. Sometimes it is described as transforming from the time domain to the frequency domain. It is very useful for analysis of time-dependent phenomena.

Spectrum Analysis & Faults Diagnosis

Machine Vibration Sources
If vibration amplitude turns to be increased at the bearing frequency, then we can determine what is wrong with the machine.

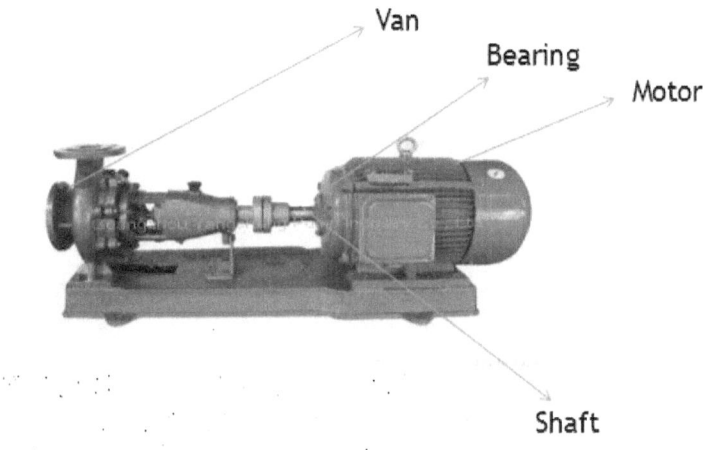

- Van
- Bearing
- Motor
- Shaft

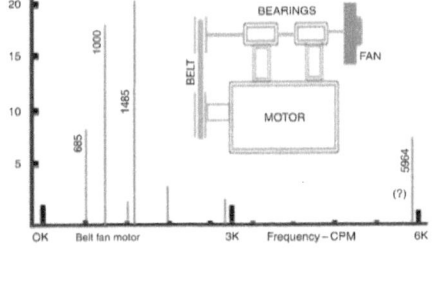

- Belt frequency 685 cpm
- Motor, 1× rpm 1485 cpm
- Fan shaft, 1× rpm 1000 cpm
- Motor, 4× rpm or electrical? 5964 cpm.

What are the Faults that Spectrum Analysis Can Tell Us About?

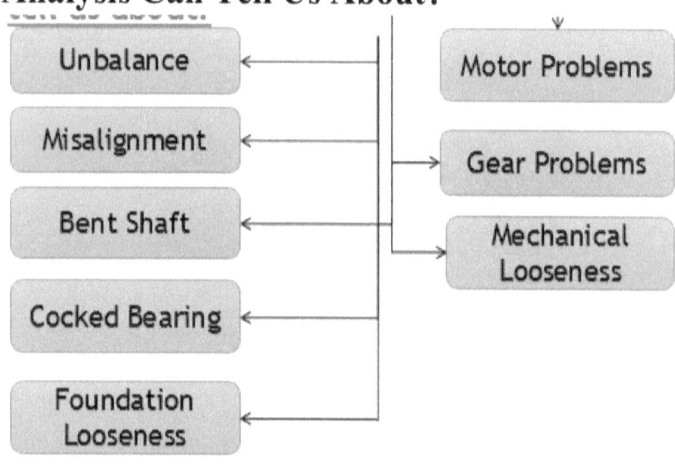

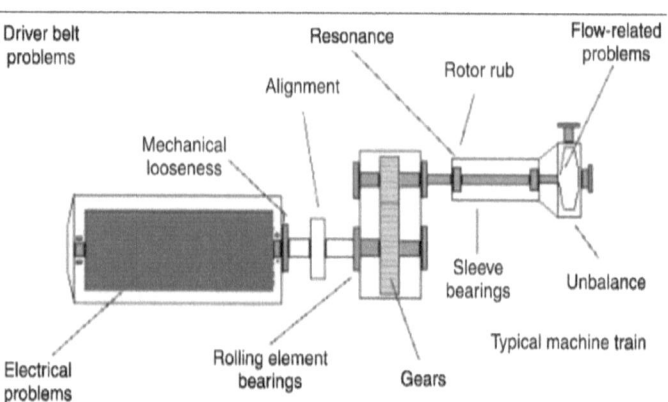

Illustrate different Machine Faults Detected by Vibration Analysis

Faults to be Detected by Spectrum Analysis

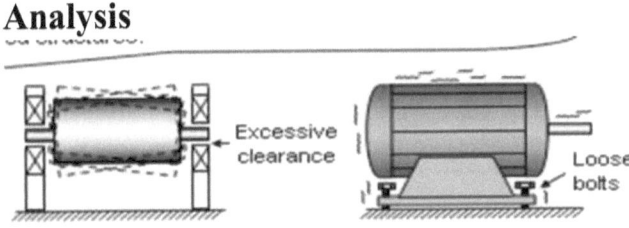

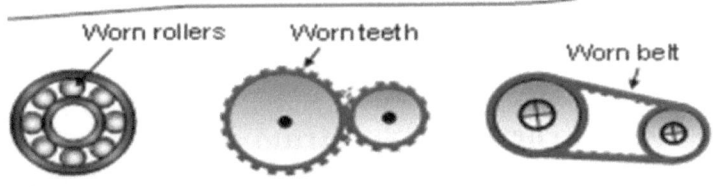

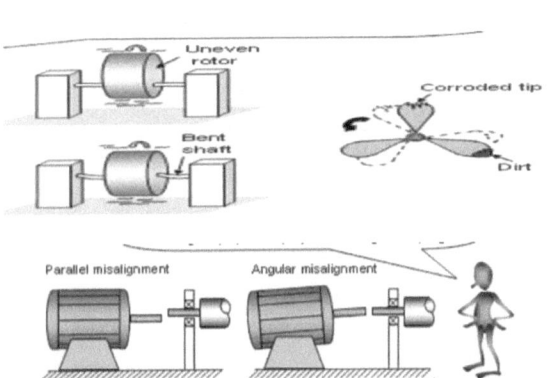

Predefined Spectrum Analysis Bands

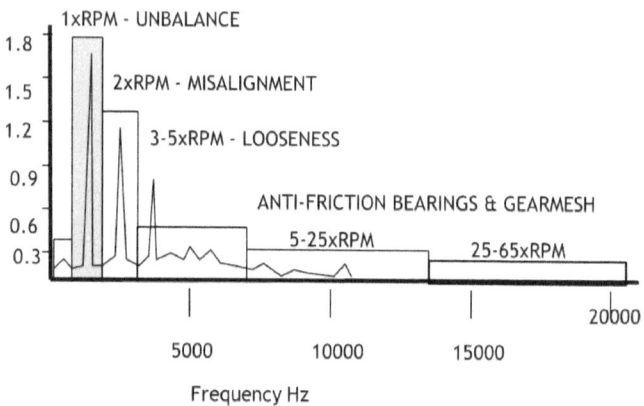

Construction of Spectrum Analysis

Frequency in terms of RPM	Most likely causes	Other possible causes and remarks
1x RPM	Unbalance	1) Eccentric journals, gears or pulleys 2) Misalignment or bent shaft- if high axial vibration 3) Resonance 4) Reciprocating forces 5) Electrical problems
2x RPM	Mechanical Looseness	1) Misalignment if high axial vibration 2) Reciprocating forces 3) Resonance 4) Bad belts if 2x RPM of belt

Frequency in terms of RPM	Most likely causes	Other possible causes and remarks
3x RPM	Misalignment	Usually a combination of misalignment and excessive axial clearances (looseness)
Less than 1x RPM	Oil whirl (less than ½ RPM)	1) Bad drive belts 2) Background vibration 3) Sub-harmonic resonance
Synchronous (A.C. Line Frequency)	Electrical Problems	Common electrical problems include broken rotor bars, eccentric rotor, unbalanced phases in poly-phase systems, unequal air gap.
2x Synch. Frequency	Torque pulses	Rare as a problem unless resonance is excited

Frequency in terms of RPM	Most likely causes	Other possible causes and remarks
High frequency (not harmonically related)	Bad anti-friction bearings	1) Bearing vibration may be unsteady - amplitude and frequency 2) Cavitations, recirculation and flow turbulence cause random, high frequency vibration. 3) Improper lubrication of journal bearings (friction excited vibration) 4) Rubbing

Frequency in terms of RPM	Most likely causes	Other possible causes and remarks
Many times RPM (harmonically related freq.)	Bad gears Aerodynamic forces Hydraulic forces Mechanical looseness Reciprocating forces	Gear teeth times RPM of bad gear Number of fan blades times RPM Number of impeller vanes times RPM May occur at 2,3,4 and sometimes higher harmonics if severe looseness

Spectrum Analysis Goals:

- Determine if a fault condition exist.
- Diagnose the fault condition.
- Determine if additional analysis is required.
- Determine severity and action required.
- Investigate root cause and provide feedback.

Spectrum Analysis Case Study: Detection of Different Failures in Rotating Machines

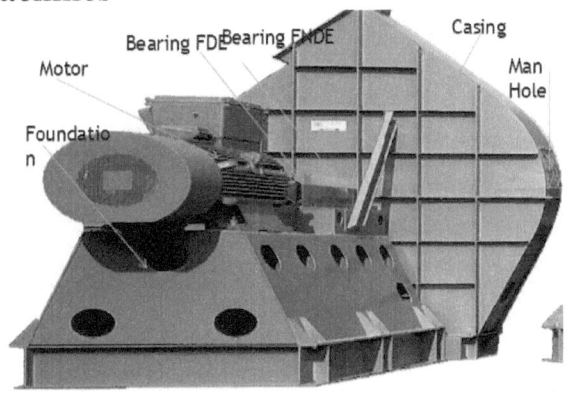

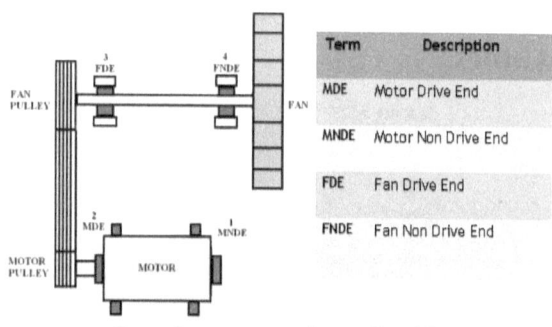

General components of centrifugal fan

1. Using Vibration Spectrum Analysis to Detect Machine Unbalance

For all types of unbalance, the FFT spectrum will show a predominant 1. Rpm frequency of vibration.

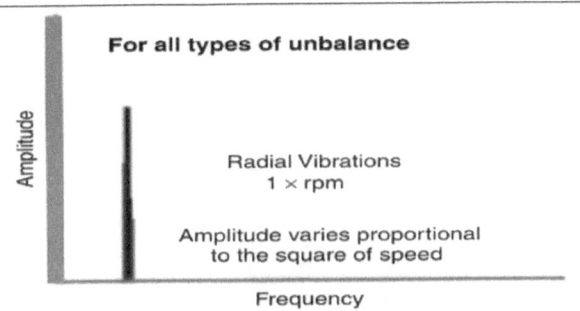

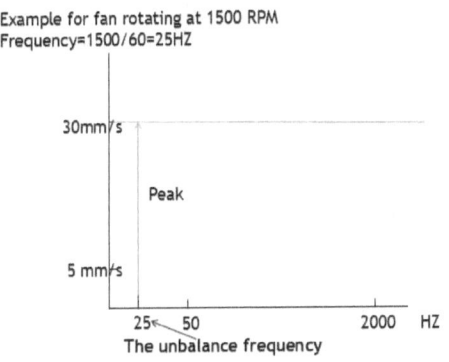

Peak=Maximum of the units being measured

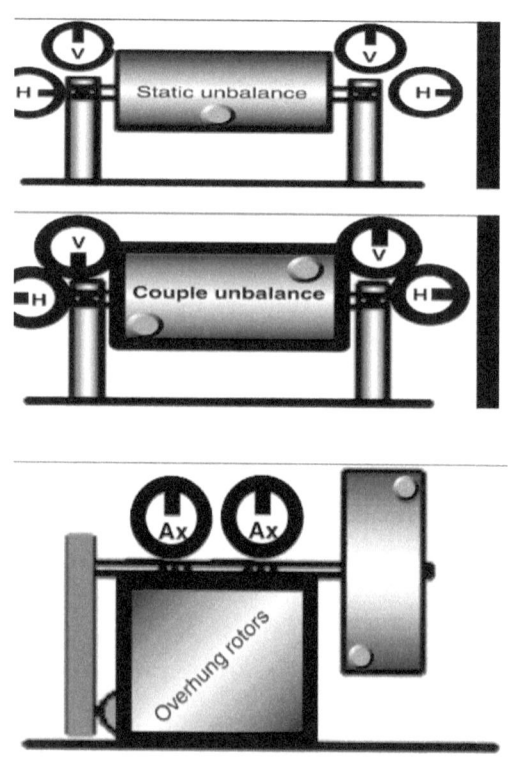

Technical Data:
Fan Type= Centrifugal fan
Flow rate= 78,000m3/h

Motor Power= 200KW
Horse power= 268HP
Motor Speed= 1500RPM
Fan Speed= 1500RPM
Transmission type: V-belts

Spare Parts Data:
FDE Bearing: 22222EK +H322
FNDE Bearing: 22222EK +H322
MDE Bearing: 6322
MNDE Bearing: 6322
Bearing Housing: SNH 522-619
Pulleys type: SPC335
V-belts type: SPC5300

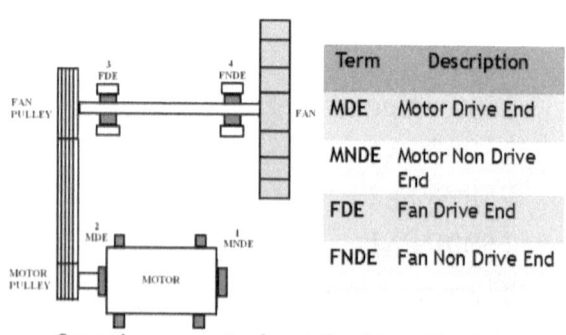

General components of centrifugal fan with v-belts transmission

Term	Description
MDE	Motor Drive End
MNDE	Motor Non Drive End
FDE	Fan Drive End
FNDE	Fan Non Drive End

Vibration Measurements

POINT	DESCRIPTION	overall values	LIMITS		
1 H	E-Motor non drive end horizontal	7.267 mm/s	x	x	
1 V	E-Motor non drive end vertical	3.611 mm/s		x	
1 A	E-Motor non drive end axial	3.276 mm/s	x	x	
2 H	E-Motor drive end horizontal	24.302 mm/s	x	x	x
2 V	E-Motor drive end vertical	3.217 mm/s		x	
2 A	E-Motor drive end axial	2.827 mm/s		x	
3 H	fan fixed bearing coupling side horizontal	35.610 mm/s	x	x	x
3 V	fan fixed bearing coupling side vertical	5.521 mm/s		x	
3 A	fan fixed bearing coupling side axial	-			
4 H	Fan free bearing fan side horizontal	6.609 mm/s		x	
4 V	Fan free bearing fan side vertical	4.941 mm/s		x	
4 A	Fan free bearing fan side axial	3.733 mm/s		x	

DANGEROUS   x x x

ALARM   x x

ACCEPTED   x

Spectrum Analysis for FDE-H

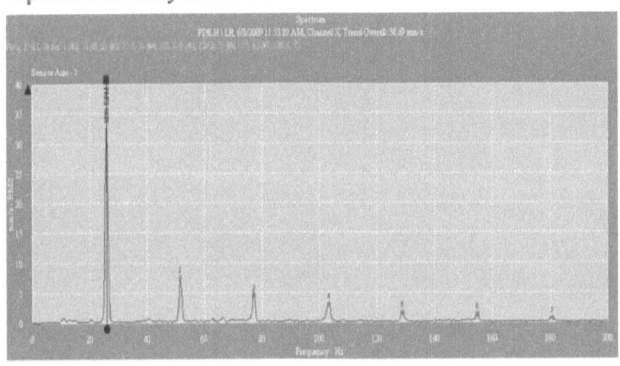

Fig. 1

Analysis
From the spectrum it is observed that the relatively high vibration in horizontal direction in the frequency range of 1X rpm can be the result of unbalance.

Recommendation
It's recommended to perform the balancing for this machine to reach the acceptable vibration limits, clean the rotor and all scales exist on the blade.

Maintenance Work Done
Sandblasting the fan blade.
Cleaning the rotor.

Balancing to the acceptable vibration limit.

2. Using Vibration Spectrum Analysis to Detect Bent Shafts

When a bent shaft is encountered, the vibrations in the radial as well as in the axial direction will be high. Axial vibrations may be higher than the radial vibrations. The FFT will normally have 1× and 2× components. If the:

Amplitude of 1x. RPM is dominant then the bend is near the shaft center.

Amplitude of 2x. RPM is dominant then the bend is near the shaft end.

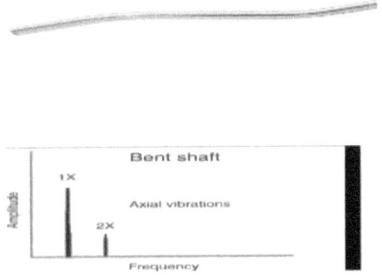

3. Using Vibration Spectrum Analysis to Detect Machine Misalignment

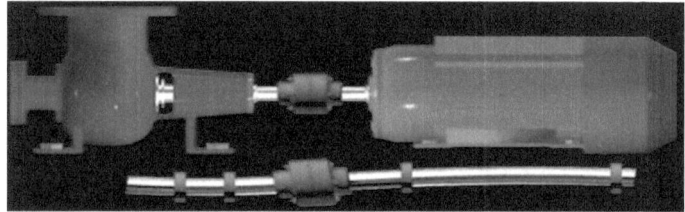

Misalignment, just like unbalance, is a major cause of machinery vibration. Some machines have been incorporated with self-aligning bearings and flexible couplings that can take quite a bit of misalignment. However, despite these, it is not uncommon to come across high vibrations due to misalignment. There are basically two types of misalignment:

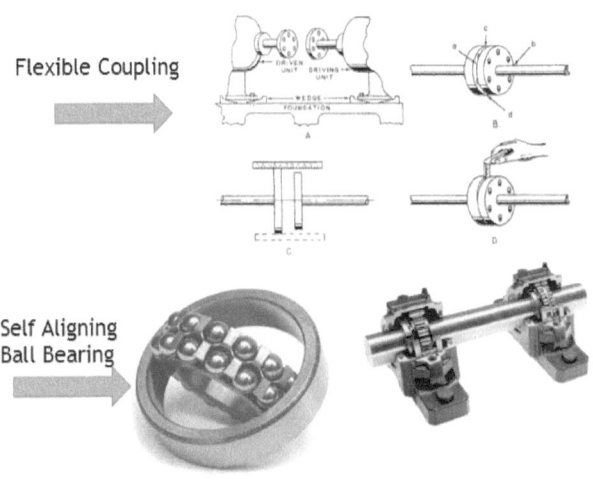

Types of misalignment

Angular Misalignment

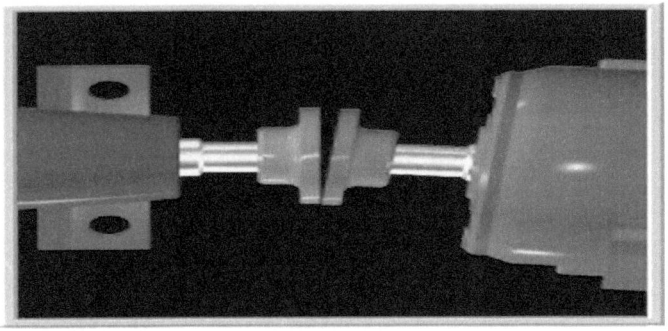

Forces are at the axial direction

1x & 2x high in spectrum, sometimes 3x

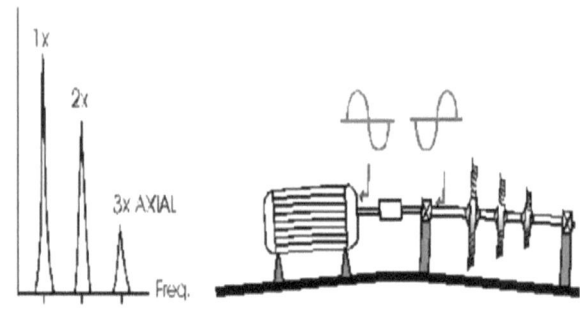

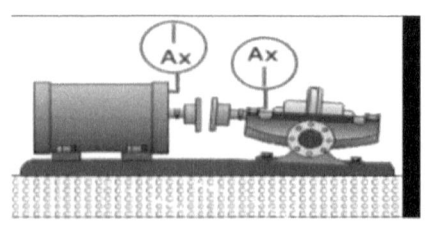

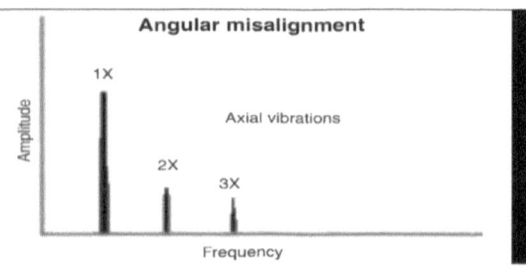

Offset (Parallel) Misalignment

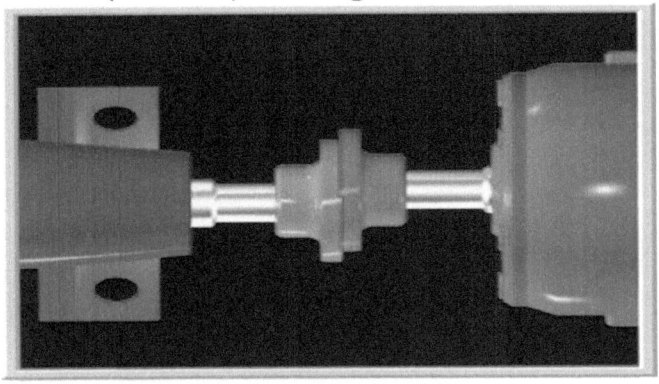

Forces are at the radial direction.

2x will be higher than 1x at radial direction.

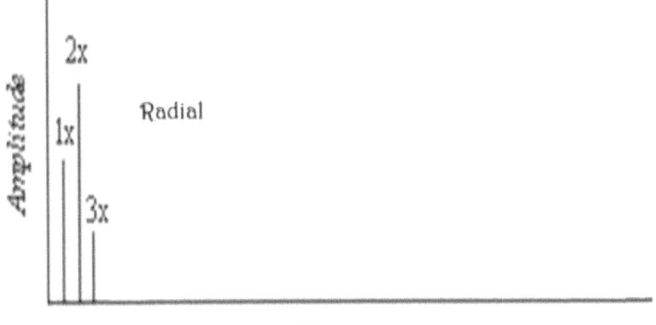

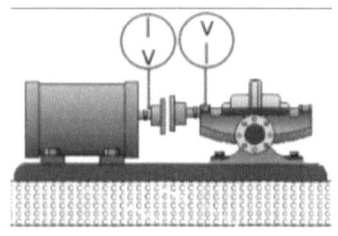

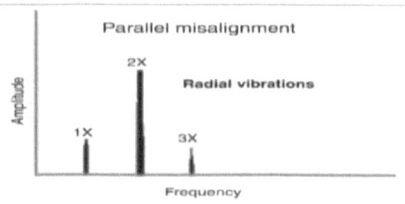

Excessive misalignment leads to several machine damages & stresses:

- Bearing Damage
- Seals Damage
- Bearing Housing Damage
- Shaft Damage
- Coupling Damage

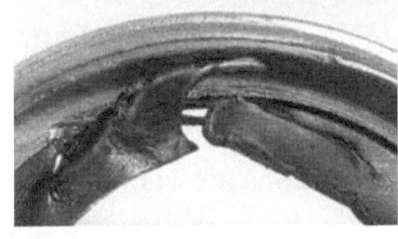
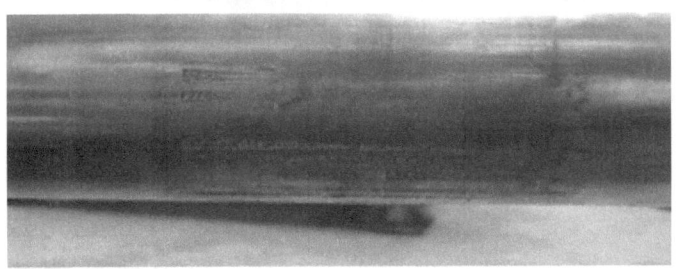

 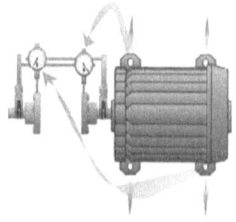

Precision Maintenance "Shaft Alignment"

Misalignment vs. bent shaft

Often, a bent shaft and dominant angular misalignment give similar FFT spectrums. The vibrations are visible in both the axial and radial vibration measurements. It is only with phase analysis that these problems can be resolved further. In a machine with a bent shaft, a phase difference will be noticed on the two bearings of the same shaft. In the case of misalignment, the phase difference is visible on bearings across the coupling.

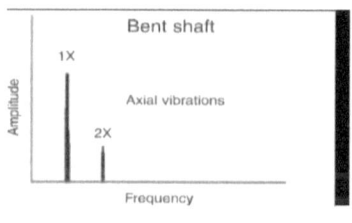 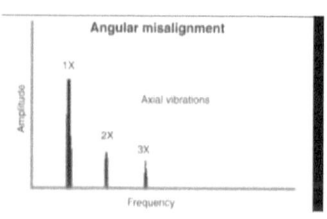

Misalignment vs. bent Shaft

Detect Transmission Belt & Sheaves Fault Condition Using Spectrum Analysis

Equipment : Electric Motor
Speed : 1500RPM
Power : 10KW
V-Belts Type : SPC5300
Pulleys Diameter: 355mm

From the tables of the v-belts specs, length is 1.2m

Belt Frequency = 3.142 (D/L) X (RPM/60)
Where D is the pulley diameter & L is the belt length
Belt Frequency=23.2HZ

Vibration Reading
M NDE Horizontal mm/sec	4.09
M NDE Vertical mm/sec	0.57
M NDE Axial mm/sec	01.99

M DE Horizontal   3.03 mm/sec
M DE Vertical   01.57 mm/sec
M DE Axial   02.12 mm/sec

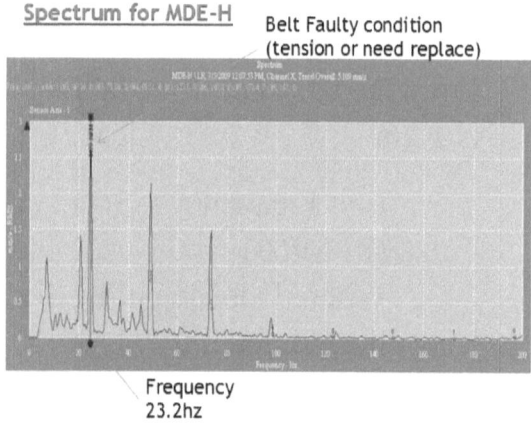

Spectrum for MDE-H — Belt Faulty condition (tension or need replace)

Frequency 23.2hz

When belts are worn, loose or mismatched, they may generate harmonics of the belt frequency. Quite often, the 2× belt frequency is dominant.

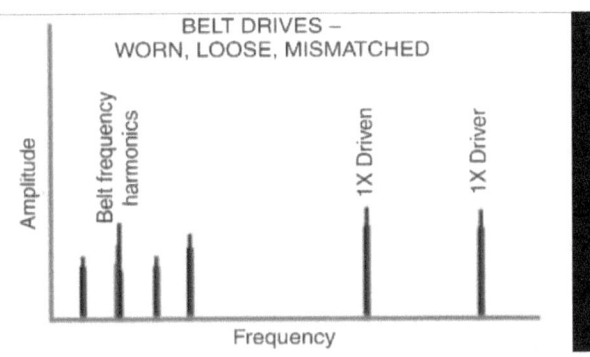

Amplitudes are normally unsteady, sometimes pulsing with either driver or driven rpm. With timing belt drives, it is useful to know that high amplitudes at the timing belt frequency indicate wear or pulley misalignment.

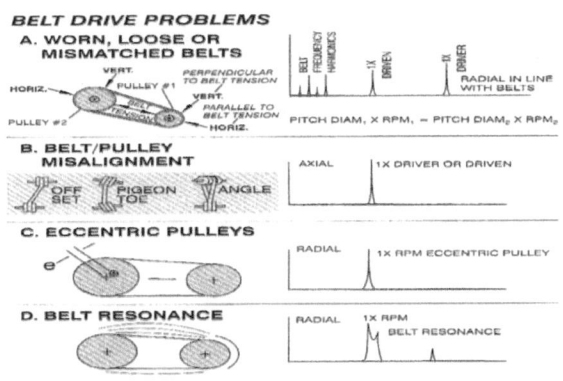

Sheaves Misalignment on Spectrum

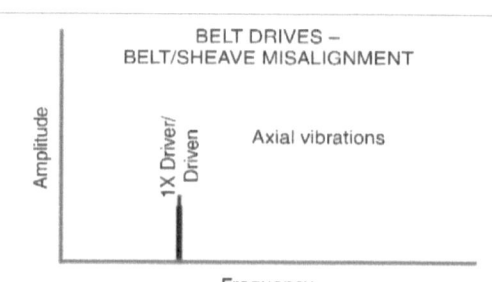

Eccentric Sheaves on Spectrum

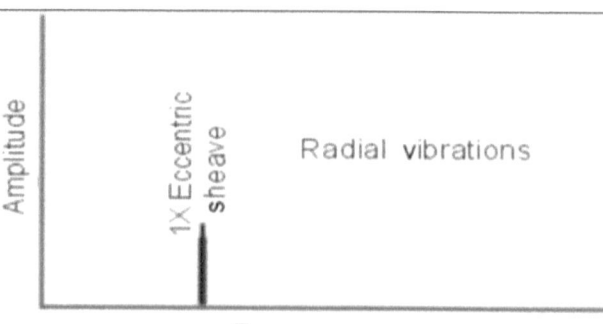

Belt/sheave misalignment

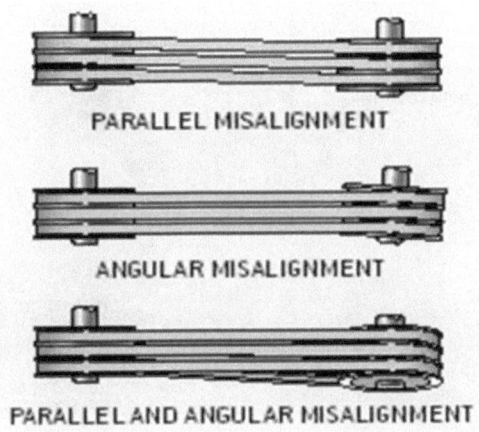

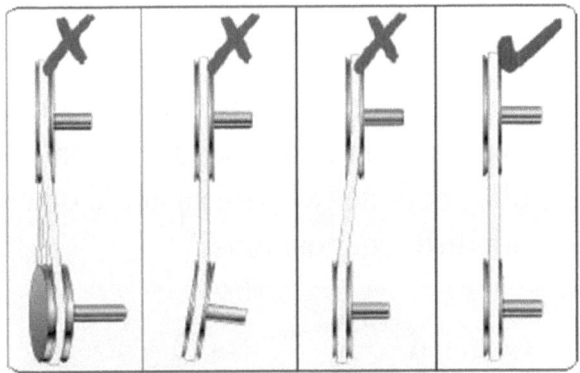

Detection of different pulleys misalignments

Taking action to align pulleys and belts
> Pulley Laser Alignment

This instrument uses two targets and a laser device. Placing the laser device on one pulley and the two targets in the other pulley with different positions.

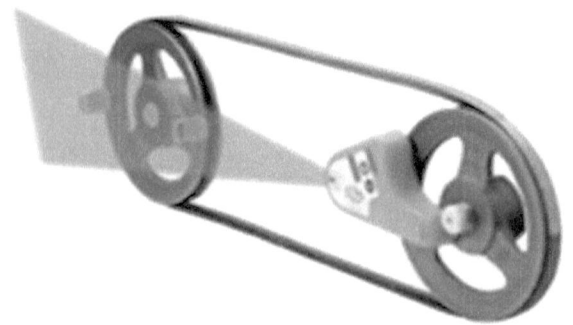

Case Study.2 "Detect electric Motor Faults by Vibration Spectrum Analysis"

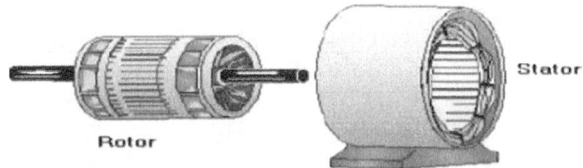

It's evident that defects in the bearings represent the widest source of failure to an induction motor, thus more focus was needed on bearings defects in particular.

Failure	Percentage
Bearings	44%
Stator	26%
Rotor	8%
Others	22%

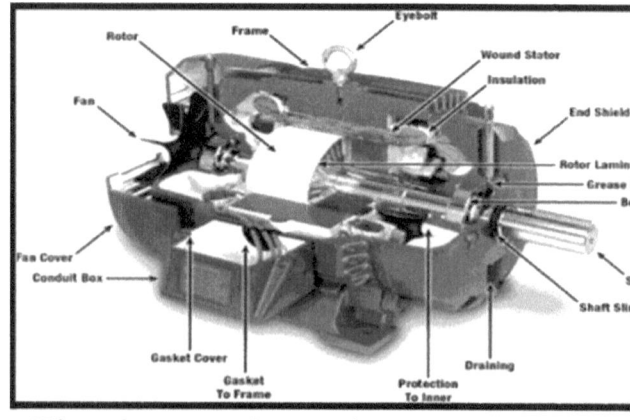

Electric motor internal components

The components of our predictive maintenance tool.

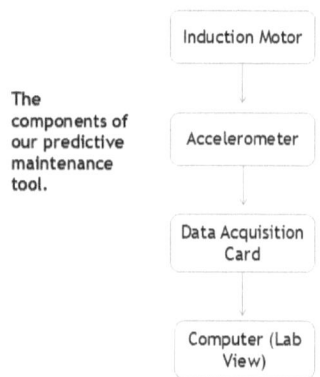

It's difficult to detect the exact reason of a bearing failure without having the vibration test applied. The following is a list of the common defect causes:

– Ordinary wear.
– Too high ambient temperature.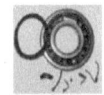
– Corrosion.
– Reduced lubrication.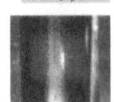
– Misalignment.
– Vibrations.
– Damage due to transport.
– Bearing currents from frequency converter drive.

1. **Detect the Nature of Bearing Failure**

When a certain defect is present on a bearing element (example of a rough defect is shown in the above figure) an increase in the vibration levels at this frequency can be noticed, and that's why frequency-domain analysis of vibration reading is usually carried out to determine the condition of motor bearings. Frequency-domain or spectral analysis of vibration signal is the most widely used approach for bearing defect detection.

Example

1500um outer race defect

Rough defect condition in a bearing

Formula to calculate the outer race defect frequency:
BPFO (Ball Pass Frequency, Outer race):

$BPFO = Nb/2(1 - Bd/PD \cos α) \times RPM$

Where:
Bd=diameter of rolling element.
PD=pitch diameter.
α=the contact angel.
Nb=no of rolling elements.
RPM= the shaft rotating frequency.

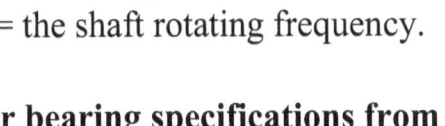

Motor bearing specifications from the tables:
Shaft rotating frequency of 24Hz
Bearing having 9 balls of diameter 8.5mm
Pitch circle diameter of 38.5mm
Contact angle a of 0.
RPM= RPM/60
Calculations=9x24/2 X (1-0.22) =84.24Hz

Bearing Health Condition

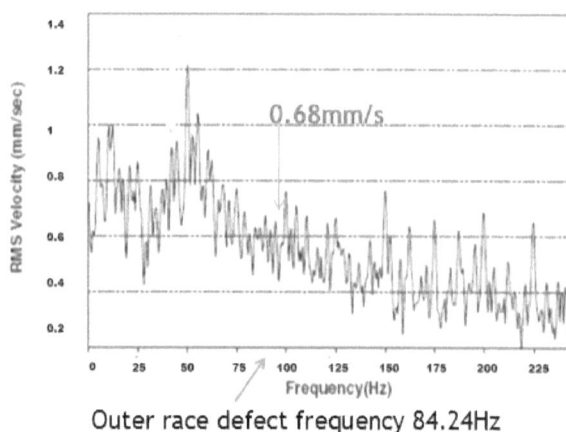

Outer race defect frequency 84.24Hz

Bearing Condition After 1 month

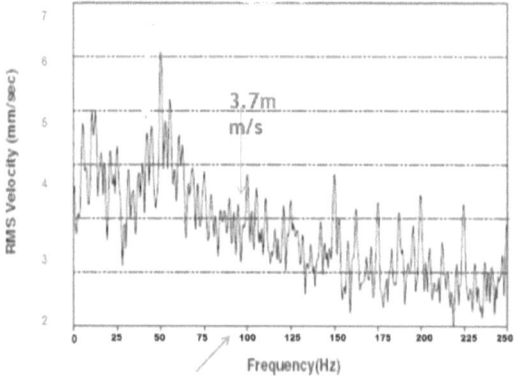

Bearing Faulty Condition

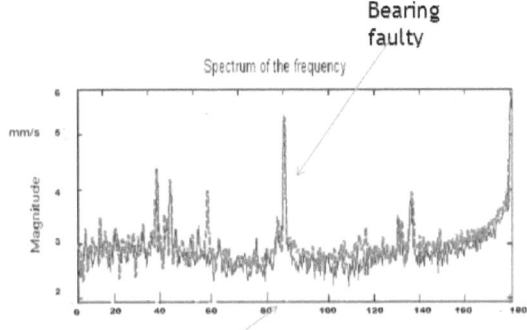

Spectrum & Vibration Analysis
BPFO defect of the bearing can be measured at 84.24hz, bearing vibration amplitude showing significant increase at the bearing outer race frequency indicating outer race defect.

Recommendation
It's highly recommended to shut down the machine to replace the defected bearing.

Common Bearing Failure Patterns

$$BPFI = \frac{Nb}{2}(1 + \frac{Bd}{Pd}\cos\theta) \times \text{rpm}$$

$$BPFO = \frac{Nb}{2}(1 - \frac{Bd}{Pd}\cos\theta) \times \text{rpm}$$

$$FTF = \frac{1}{2}(1 - \frac{Bd}{Pd}\cos\theta) \times \text{rpm}$$

$$BSF = \frac{Pd}{2Bd}\left[1 - \left(\frac{Bd}{Pd}\right)^2(\cos\theta)^2\right] \times \text{rpm}$$

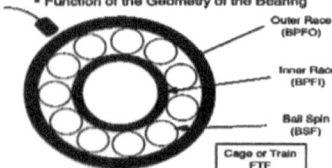

Nb = Number of Balls or Rollers
Bd = Ball / Roller diameter (inch or mm)
Pd = Bearing pitch diameter (inch or mm)
θ = Contact angle in degrees

BPFI = Ball pass frequency – Inner
BPFO = Ball pass frequency – Outer
FTF = Fundamental train frequency (Cage)
BSF = Ball spin frequency (rolling element)

2. Detection of Different Electric Problems

The following are some terms that will be required to understand vibrations due to electrical problems:

F_L = electrical line frequency (50/60 Hz)

F_s = slip frequency = $\dfrac{2 \times F_L}{P}$ − rpm

F_p = pole pass frequency = $F_s \times P$

P = number of poles.

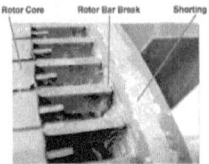

Rotor Problems
- Broken rotor bars
- Open or shorted rotor windings
- Bowed rotor
- Eccentric rotor

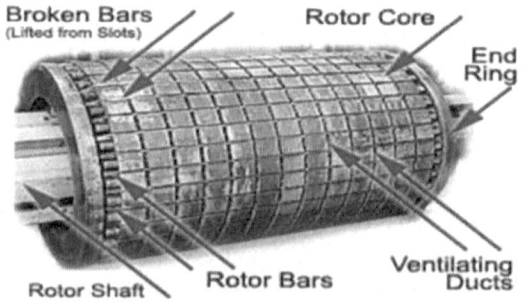

Defect rotor bars

A. Rotor Defects
Broken rotor bars
Eccentric rotor

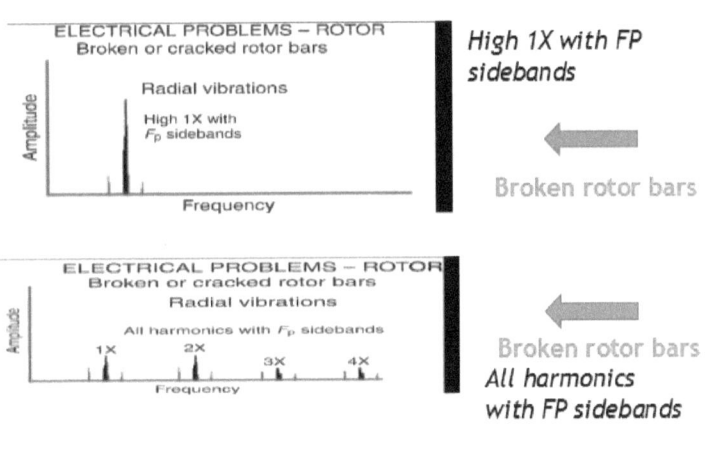

*High 1X with FP sideband*s

Broken rotor bars

Broken rotor bars
All harmonics with FP sidebands

Example:
Motor Speed Synchronous =1800RPM
Motor Speed Actual =1770RPM
No of poles=4
FL=60hz
FP= 2xFL/P-RPM*P=2hz

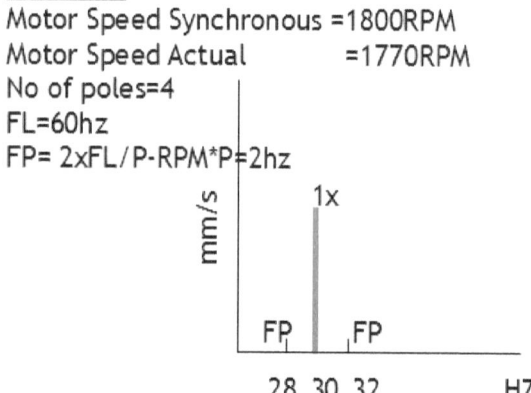

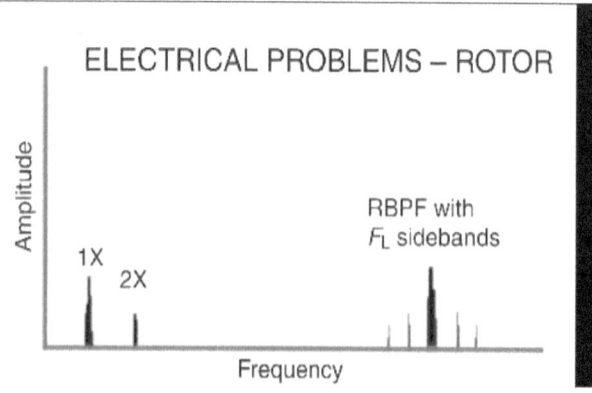

Rotor bar pass frequency, RBPF = number of rotor bars. RPM/60

B. Eccentric Rotor

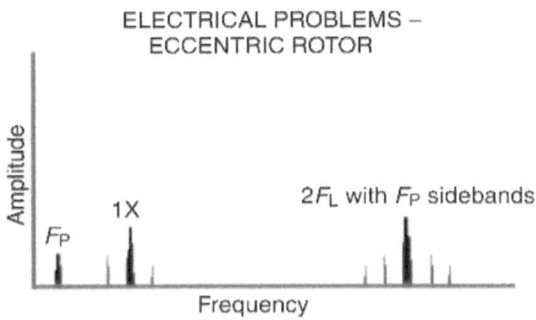

C. Stator Defects
Eccentricity
Short lamination
Loose iron

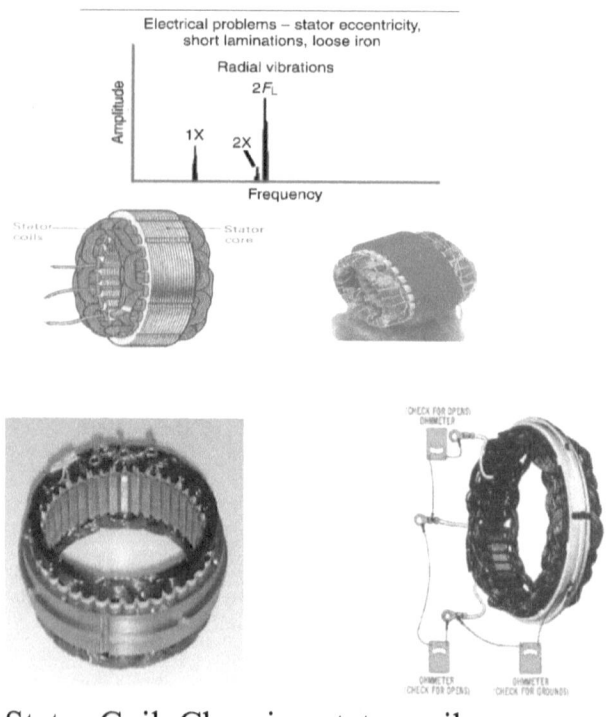

Stator Coil, Charging stator coils

D. Phasing Problem (loose connector)
Often occurs due to broken connectors

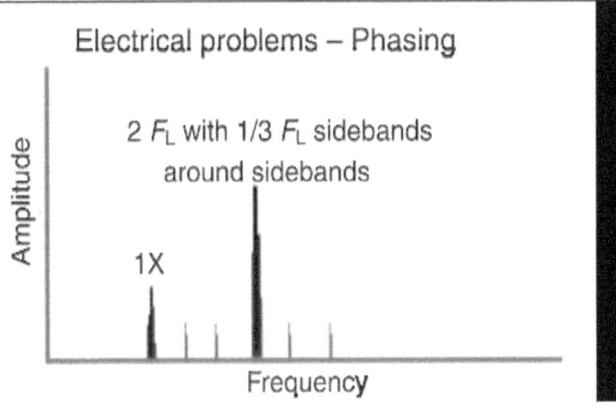

E. Synchronous Motors (Loose stator coils)

CPF = number of stator coils.
RPM/60

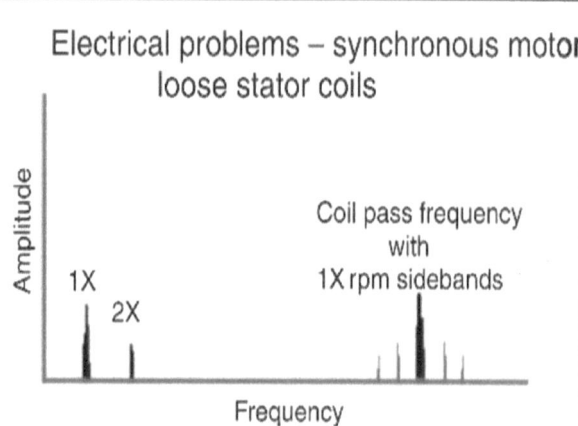

F. DC motor problems

Possible reasons:
Broken field windings.
Bad SCRs.
Loose connections.
Loose or blown fuses.
Shorted control cards.

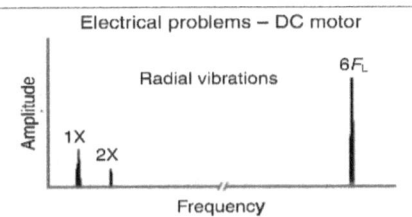

Case study. 3 Using Vibration Spectrum Analysis to Detect Machine Looseness

If we consider any rotating machine, mechanical looseness can occur at three locations:
1. Internal assembly looseness
2. Looseness at machine to base plate interface
3. Structure looseness.

I. Internal Assembly Looseness

This category of looseness could be between a bearing liner in its cap, a sleeve or rolling element bearing, or an impeller on a shaft. It is normally caused by an improper fit between component parts, which will produce many harmonics in the FFT due to the nonlinear response of the loose parts to the exciting forces from the rotor.

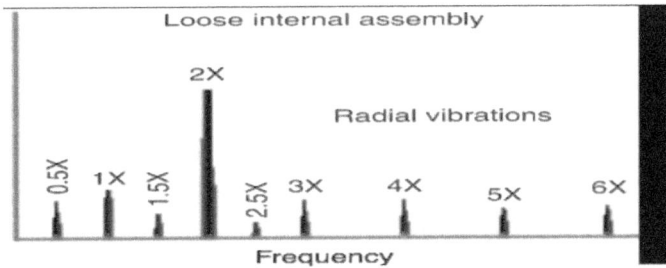

II. Looseness Between Machine to Base Plate

This problem is associated with loose pillow-block bolts, cracks in the frame structure or the bearing pedestal.

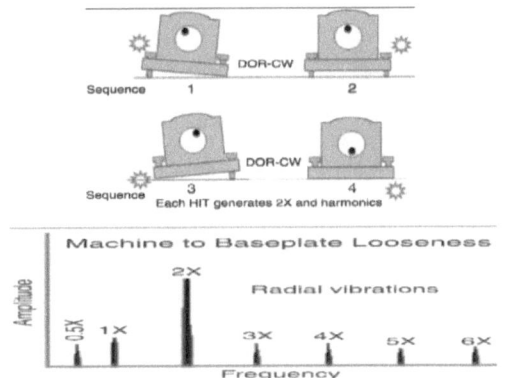

III. Structure Looseness

This type of looseness is caused by structural looseness or weaknesses in the machine's feet, baseplate or foundation.

Soft foot
Structure looseness

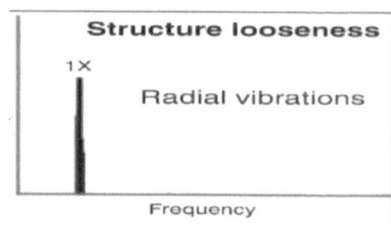

When the soft foot condition is suspected, an easy test to confirm for it is to loosen each bolt, one at a time, and see if this brings about significant changes in the vibration. In this case, it might be necessary to re-machine the base or install shims to eliminate the distortion when the mounting bolts are tightened again.

Case study.4: Detection of Bearing Looseness

Equipment data:

Machine Description: Centrifugal de-gasing fan
Horse Power: 350HP
Fan speed= 992RPM
Flow rate: 300,000 m3/h
No of bearing: 6 bearings
Bearing designation: 22222EK

General Components of the Fan

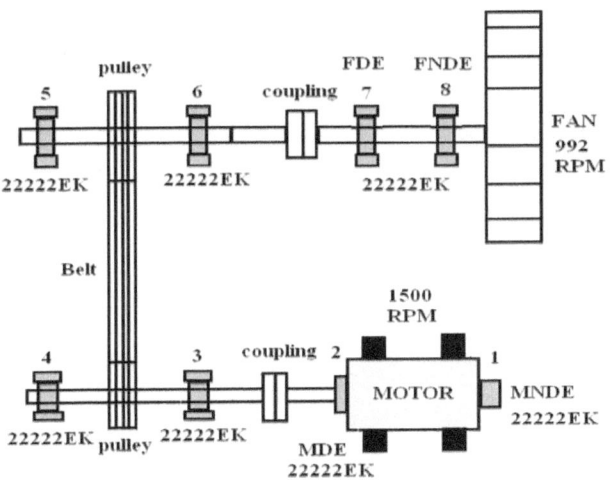

de-gasing fan15B802

Vibration Reading:

Bearing 3-H	= 5.965
Bearing 3-V	= 8.632
Bearing 4-H	= 16.042
Bearing 4-V	= 12.828
Bearing 5-H	= 5.83
Bearing 5-V	= 5.914
Bearing 6-H	= 7.812
Bearing 6-V	= 6.505
FDE-H	= 6.512
FDE-V	= 11.878
FNDE-H	= 4.805
FNDE-V	= 17.869

Spectrum for bearings 4 & 5

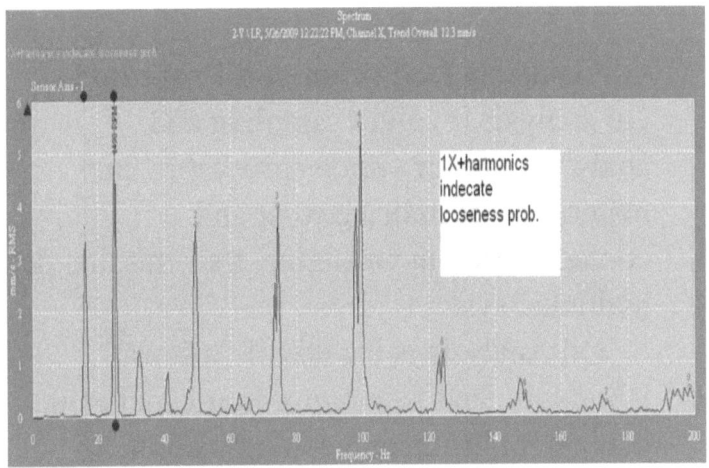

Spectrum Analysis:
Check Bearing no. (4) for fixation problem with the base. (Retighten bearing housing bolts & check bearings internal clearance).

Check Bearing no. (5) For fixation problem with the base. (Retighten bearing housing bolts & check bearings internal clearance).

Maintenance work to be performed:
1-Retightining of bearing bolts & housing.
2-Monitor bearing condition with some other methods (use ultrasonic)

Oil Analysis

Oil Analysis Definition and Procedures

Oil analysis involves sampling and analyzing oil for various properties and materials that indicate wear and contamination in an engine, transmission or hydraulic system.

Oil Analysis is the use of various laboratory tests to monitor lubricant health, equipment health and contamination.

Oil analysis (OA) is the sampling and laboratory analysis of a lubricant's properties, suspended contaminants, and wear debris. OA is performed during routine preventive maintenance to provide meaningful and accurate information on lubricant and machine condition.

Oil Analysis as a Part of Predictive Maintenance Programs PdM

Oil analysis technique utilize 15% of condition monitoring programs. Oil analysis is a long-term program that, where relevant, can eventually be more predictive than any of the other technologies. It can take years for a plant's oil program to reach this level of sophistication and effectiveness.

Analytical techniques performed on oil samples can be classified in two categories:
Used Oil Analysis
Wear Particles Analysis

Used oil analysis determines the condition of the lubricant itself, determines the quality of the lubricant, and checks its suitability for continued use.

Oil Analysis procedures

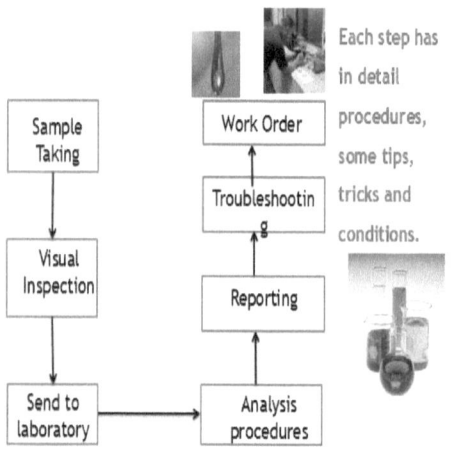

Sample Taking → Visual Inspection → Send to laboratory → Analysis procedures → Reporting → Troubleshooting → Work Order

Each step has in detail procedures, some tips, tricks and conditions.

Terminologies Involve in Lubricant Systems

These terminologies will be used frequently in any oil analysis program or application.

Flash Point: the point at which the oil will

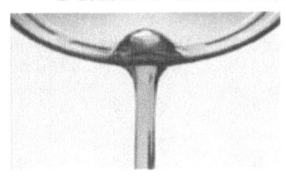 

be turned into vapor or begin to vaporize.

Pour Point: the lowest temperature at which the oil will flow.

TBN: is the total base number, illustrate the no of he active additives left in the sample of oil to neutralize the acids. By comparing the TBN of a used oil to the TBN of the same oil in virgin condition, the user can determine how much reserve additive the oil has left to neutralize acids. The lower the TBN reading, the less active additive the oil has left.

The Viscosity Index is a measure of how much the oil's viscosity changes as temperature changes. A higher viscosity index indicates the viscosity changes less with temperature than a lower viscosity index.

TAN: is the total acid number, present how this oil is getting oxidized.

Role and Types of Different Oil Additives
Oil additives are chemical compounds that improve the lubricant performance of base oil. They vary from an oil to another based on the oil function. For example, car engine oil may contain viscosity modifiers to improve oil performance at different temperatures. While electric transformers oil may contain specific type of additives to improve oil insulation. It's really based on the oil function inside the equipment. Engine oils are used to protect from wears and tears. Transformer oils are used for isolation and cooling purposes. So what you basically look for in each oil type is totally different.

Detergent additives, dating back to the early 1930, are used to clean and neutralize oil impurities which would normally cause deposits (oil sludge) on vital engine parts.

Friction modifiers, like molybdenum disulfide, are used for increasing fuel economy by reducing friction between moving parts. Friction modifiers alter the lubricity of the base oil. While oil was used historically.

Anti-wear additives or wear inhibiting additives cause a film to surround metal parts, helping to keep them separated. Zinc Dialkyl Dithio Phosphate (ZDDP) is a popular anti-wear additive, the use of which has been restricted thanks to potential damage to catalytic converters forced upon automakers by government regulation.

Pour point depressants improve the oil's ability to flow at lower temperatures.

Anti-foam agents inhibit the production of air bubbles and foam in the oil which can cause a loss of lubrication, pitting, and corrosion where entrained air contacts metal surfaces.

Corrosion or rust inhibiting additives retard the oxidation of metal inside an engine.

Antioxidant additives retard the decomposition of the stock oil.

Viscosity modifiers make an oil's viscosity higher at elevated temperatures, improving its viscosity index (VI). This combats the tendency of the oil to become thin at high temperature. The advantage of using less viscous oil with a VI improver is that it will have improved low temperature fluidity as well as being viscous enough to lubricate at operating temperature. Most multi-grade oils have viscosity modifiers. Some synthetic oils are engineered to meet multi-grade specifications without them.

Seal conditioners cause gaskets and seals to swell so that the oil cannot leak by.

Metal deactivators create a film on metal surfaces to prevent the metal from causing the oil to be oxidized.

Extreme pressure agents bond to metal surfaces, keeping them from touching even at high pressure.

Dispersants keep contaminants (e.g. soot) suspended in the oil to prevent them from coagulating.

Wax crystal modifiers are DE waxing aids that improve the ability of oil filters to separate wax from oil. This type of additive has applications in the refining and transport of oil, but not for lubricant formulation.

Wear metals from friction are unintentional oil additives, but most large metal particles and impurities are removed in situ using either magnets or oil filters made for this purpose.

Most common elements found in the additives:

Barium (Ba), detergent or dispersant additive.

Boron (B), extreme-pressure additive.

Calcium (Ca), detergent or dispersant additive.

Copper (Cu), anti-wear additive.

Lead (Pb), anti-wear additive.

Magnesium (Mg), detergent or dispersant additive.

Molybdenum (Mo), friction modifier.

Phosphorus (P), corrosion inhibitor, anti-wear additive.

Silicon (Si), anti-foaming additive.

Sodium (Na), detergent or dispersant additive.

Zinc (Zn), anti-wear or anti-oxidant additive.

What can oil analysis tells you?

1. **Lubricant Health**
- Monitor changes in lubricant properties, determine the suitability for continued use:
- Additives metals analysis by ICP, Magnesium, Calcium, Silicon, Phosphorus, Zinc, Barium.
- Viscosity-Resistance to flow at temperature.
- Total Acid Number-Detect presence of acids.

- Total Base Number-Measure the ability to neutralize acids

2. **Equipment Health**
- Detect ingression from external source
- Is filter change required-high particles count
- Is repair necessary-coolant leakage

3. **Elemental analysis by Induction Couple plasma ICP**
- Water
- Fuel Dilution
- Soot

4. **Determine effectiveness of maintenance strategy:**
- Reactive
- Preventive
- Predictive

Basic Tests

Component Type	Elemental Analysis	Viscosity	Water	Acid NO	Oxidation	Particle Count
Engine	x	x	x		x	
Hydraulics	x	x	x	x	x	x
Gearbox	x	x	x	x	x	
Compressors	x	x	x	x	x	x
Turbines	x	x	x	x	x	x

Particle count is crucial to hydraulic systems, compressors, turbines, robotics and injection molding machines. When evaluating particle count data, particulate contamination has an immediate effect on the system. Clearance-size particles can cause slow response, spool jamming, surface erosion, solenoid burnout and may cause safety systems to fail.

Why TAN test shouldn't be used when sampling engine oils?

Engines produce carbon and other combustion components that will mix with oil and increase its acidity. Testing acidity in this case can be miss-leading. It's better to perform TBN to test how this oil is losing its basicity over time. There is a minimum level of TBN for each oil type.

**Different Types of Oil Elemental Tests and Techniques Used
Elemental Analysis by ICP (Inductively Coupled Plasma)**

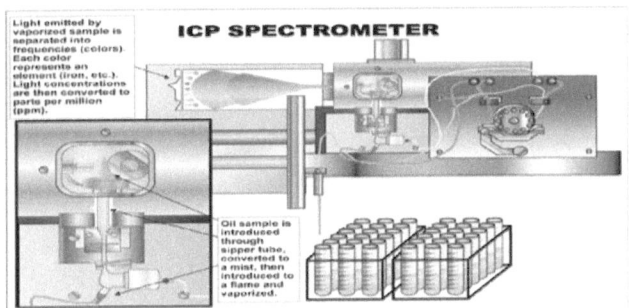

Test Methods: using a Rot rode Emission Spectrometer or an Inductively Coupled Plasma Spectrometer.

The Rot rode Spectrometer has a particle size detection limitation of between 3µ and 10µ (depending on the particular metal in question and the amount of surface oxidation on the particle surface) compared to the 0.5µ - 2µ limitation of the ICP.

Notice some metals can be both additives and contaminants, such as Calcium, or wear metals and additives, such as Zinc. The following table show so.

Elements are classified into:

Wear	Additives	Contaminant
Iron	Silicon	Silicon
Lead	Boron	Boron
Copper	Copper	Phosphorous
Tin	Sodium	Potassium
Aluminum	Phosphorous	Calcium
Chromium	Zinc	Magnesium
Nickel	Calcium	Vanadium
Silver	Magnesium	
Titanium	Molybdenum	
Antimony	Antimony	
Zinc	Potassium	

Viscosity Test

✓ Most important lubricant property (lubrication selection begins with viscosity calculation.
✓ Measure fluid resistance to flow under gravity.
✓ Relative to fluid density, thickness.
✓ Affects fluid ability to lubricate under different operating conditions.

Water Tests

Test is performed by several methods:

1. Crackle
✓ Hot plate test, accurate up to 0.5%
✓ Very subjective > 0.5%

✓ Estimated
2. **Karl Fischer ASTM D1744**
   Good test for turbine compressors, hydraulics, and some gear oils. Reported in % or ppm.
3. **Infrared**
   Fair accuracy unless contaminated with other
   Factors: fuel, soot, glycol.
4. **Distillation**
   Very good test, usually cost prohibitive.

Water by Crackle %

Result is measured by placing a few drops of oil on a hot plate that is heated to 150° C. Positive water will bubble and crackle. Test is an estimate of % by volume and POLARIS Laboratories™ only reports an estimate up to 0.5%. The test is very subjective beyond 0.5%.

Test Limitation:

Accurate readings are difficult on emulsified fluids. Test is subjective and requires a minimum of 0.1% or 1000ppm to indicate a positive result. Should only be considered as a screening device for industrial hydraulics, gear boxes, compressors, bearing systems or turbines. Karl Fischer water should be performed on these unit types.

Oxidation Test
Oxidation measures the breakdown of a lubricant due to age and operating conditions. It prevents additives from performing properly, promotes the formation of acids and increases viscosity.

Nitration indicates excessive "blow-by" from cylinder walls and/or compression rings. It also indicates the presence of nitric acid, which speeds up oxidation. Too much disparity between oxidation and nitration can point to air to fuel ratio problems. As oxidation / nitration increases, so will total acid number and viscosity, while total base number will begin to decrease. Nitration is primarily a problem in natural gas engines.

Acid/Base Number Test

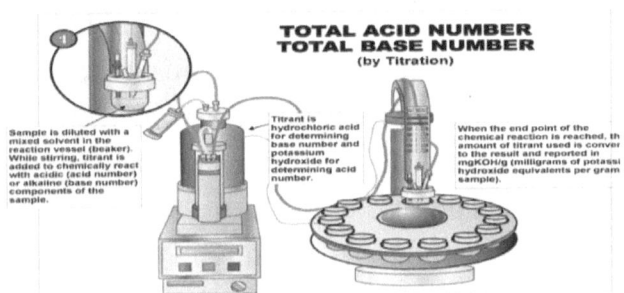

Oxidation is the most predominant reaction of a lubricant in service. It is responsible for numerous lubricant problems including:

- ✓ Viscosity increase
- ✓ Varnish, sludge and sediment formation
- ✓ Additive depletion
- ✓ Base oil breakdown
- ✓ Filter plugging
- ✓ Loss in foam control
- ✓ Acid number (TAN) increase
- ✓ Rust formation and corrosion

Oxidation is always an indication of the oil age when compared to the new oil.

Particles Count Test

Particle Quantifier (Ferrous Density) exposes a lubricant to a magnetic field. The presence of any ferrous metal causes a distortion in the field, which is represented as the PQ Index, an arbitrary unit of measurement that correlates well with DR Ferro large. Although PQ does not provide a ratio of small to large ferrous particles, if the PQ Index is smaller than iron parts per million (ppm) by ICP, it's unlikely there are any particles larger than 10 microns present. If the PQ Index increases dramatically while the ICPs iron parts per million (ppm) remains consistent or goes down, larger ferrous particles are being generated. Analytical Ferrography should then be used to qualify the type of wear occurring.

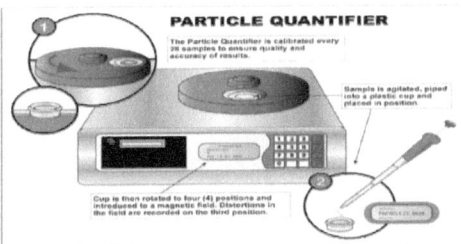

Wear Particle Analysis

Determines the mechanical condition of machine components that are lubricated. Eg. This test allows identifying which sprocket or gear tooth are faulty inside a gearbox by analyzing metal contents to identify where the failure is coming from.

Oil Analysis:

Sample
↓
Survey
↓
Results
↓
Improvement

Case Study.1: Analyzing Engine Oil

Data Required To Perform Oil Analysis

Oil Grade: Eg. 0W-40, 5W-30
Oil Type:
Equipment Type:
Oil Specifications:
Viscosity@40degC
Viscosity@100degC
Total Base Number (TBN)
Total Acid Number (TAN)
Flash Point
Oil Manufacturer: Asmoil
Oil Grade   : Synthetic Motor Oil SAE 5W-30
Oil type   : Engine Oil
Filter type   : Full Flow

Oil Analysis Report:

Parameters	Standard acc to SAE for engine grade 30	1-1-2009	1-1-2010
Viscosity@40C	90-110cSt	100	98
Viscoisty@100C	9.3-12.5cSt	11.5	8.6
Fuel Content %	0	0	5
Water Content %	0	0	2
TAN	....	....	....
TBN	6-8	7	3.82
Flash Point rate for Asmoil 5W-30 grade	375°F<	385	375

Rermark: cSt=Centi Stoke

Parameters	Standard acc to SAE for engine grade 30	1-1-2009	1-1-2010
Viscosity@40C	90-110	100	98
Viscoisty@100C	9.3-12.5	11.5	8.6
Fuel Content %	0	0	5
Water Content %	0	0	2
TAN	....	....	....
TBN	6-8	7	3.82
Flash Point rate for Asmoil 5W-30 grade	375°F<	385	375

≻Green color **indicate good condition**, yellow color **indicate alarm level**, red color indicate that oil must be changed.

Recommendations based on the result:

- Viscosity has been broken down at 100degC, oil should be changed due to reduction in viscosity index number

which will reduce the friction resistivity of the oil.
- Oil is contamination with fuel indicating leakage in the piston rings.
- Oil has water contents indicating failure head gasket or worn cylinder head, check for compression ratio of your engine.
- TBN is below the oil standard, this indicate wear, oil must be changed.
- Flash point is below the oil number, oil can be vaporized easily.

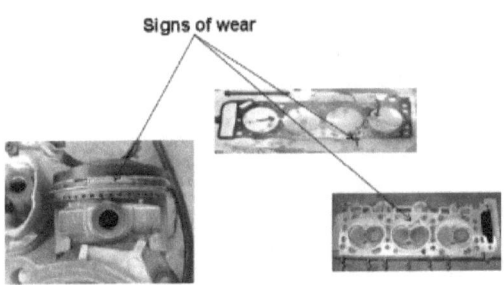

Signs of wear

Sampling Methods:

The frequency of sample analysis from your equipment depends on the machine type, machine application and condition, operating environment and other variables. For example, many machines that operate in harsh environments, such as heavy equipment in mining or construction, require short oil sampling intervals - every 100 to 300 operating hours.

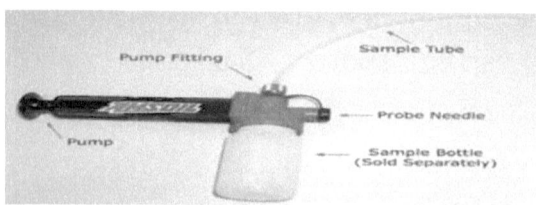

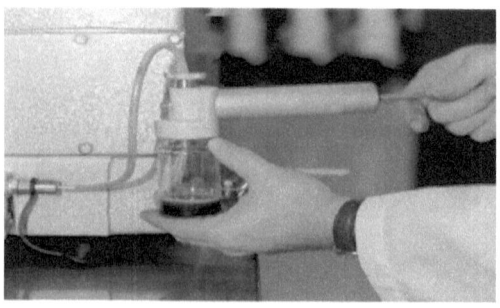

Oil sample taken can be sent to the laboratory for testing or to be tested on field using a special analyzer.

Results reported in 2 to 4 days after the lab receives the sample.

Benefits of Oil Analysis:
- ✓ Indicate how the equipment was used.
- ✓ Illustrate what condition equipment is in.
- ✓ Indicate the presence of contaminants.
- ✓ Tell if you are using the proper lubricant.

Targets from the Oil Analysis Program:

- ✓ Eliminating too-frequent oil changes.
- ✓ Reduce the cost for oil and servicing.
- ✓ Detect hidden wears.
- ✓ Detect machine health condition& evaluate it.
- ✓ Determine root causes of failures.

Vibration VS Oil Analysis

Vibration VS Oil Analysis

Condition	Oil program	Vibration program	Correlation
Water in oil	Strong	Not applicable	Water can lead to a rapid failure. It is unlikely that a random monthly vibration scan would detect the abnormality.
Greased bearings	Mixed	Strong	It makes economic sense to rely on vibration monitoring for routine greased bearing analysis. Many lube labs do not have enough experience with greased bearings to provide reliable Information.

Condition	Oil program	Vibration program	Correlation
Greased motors-operated valves	Mixed	Weak	Actuators are important machinery in the nuclear industry. Grease samples can be readily tested, but it can be difficult to obtain a representative sample. It can be hard to find these valves operating, making it difficult to monitor with vibration Techniques.
Shaft cracks	Not applicable	Strong	Vibration analysis can be very Strong effective to monitor a cracked Shaft.

Condition	Oil program	Vibration program	Correlation
Lubricant condition monitoring	Strong	Not applicable	The lubricant can be a significant cause of failure.
Resonance	Not applicable	Strong	Vibration program can detect a resonance condition. Lube analysis will eventually see the effect.
Root Cause Failure Analysis	Strong	Strong	Best when both programs work together.

Condition	Oil program	Vibration program	Correlation
Gear wears	Strong	Strong	Vibration techniques can link a defect to a particular gear. Lube analysis can predict the type of failure mode.
Alignment	Not applicable	Strong	Vibration program can detect a misalignment condition. Lube analysis will eventually see the effect of increased/improper bearing load.

Oil Analysis Can be used with anything use oil such as:

- ✓ All gearboxes for (trucks, cars, stirrers, turbines, industrial equipment…etc).
- ✓ All engines that use oil (cars, trucks, heavy equipment,….etc).
- ✓ All hydraulic systems.

Case Study.2: Turbine Oil Condition Monitoring

Types of tests performed on turbines according to the American Society for Testing Materials

Viscosity ASTM D445

Viscosity is the most important property of any lubricant.

Viscosity is defined as the resistance to flow of oil at a given temperature and is measured via the ATSM D445 protocol. As it relates to turbine oils, significant changes to viscosity usually indicate that the oil has become contaminated with another oil. In very severe cases, viscosity will increase as a result of excessive oxidation. Thermal cracking (from excessive heat) of the base oil can cause the viscosity to decrease.

Oxidation Stability by Rotary Pressure Vessel Oxidation Test (ASTM D2272)

Rotary Pressure Vessel Oxidation Test (RPVOT and formerly known as RBOT) is a measure of remaining oxidation life when compared to new oil. The test is not intended to draw comparisons between two different new oils or oils of different chemistries. In fact, oils with very high new oil RPVOT values have been seen to have the short test life in laboratory rig testing. ASTM D4378 defines 25 percent of the new oil RPVOT value as the lower limit. When the oil is approaching the 25 percent of new oil value in conjunction with an increasing TAN, ASTM D4378 recommends that plans should be made to replace the charge of oil.

Total Acid Number (ASTM D974)

Total Acid Number (TAN) is the measure of the oil's acidity and is measured by titrating the oil with a base material (KOH) and determining the amount of base required to neutralize the acids in the oil. The results are reported as mg KOH/g of the oil being tested.

TAN measures the acidic by products formed during the oxidation process. ASTM D4378 (In Service Monitoring of Mineral Turbine Oils for Steam and Gas Turbines) recommends that a 0.3 to 0.4 mg KOH/g rise above the new oil value as the warning limit. Any significant change in TAN should be investigated as the acids in the oil can cause corrosion of bearing surfaces that result in irreparable damage. However, care should be taken in reacting to a single high TAN result. The TAN test is not a precise method (+/- 40 percent by ASTM Standard) and is subject to variability of operators. Poor maintenance of the buffer solution or electrodes used in the titration can also yield false results.

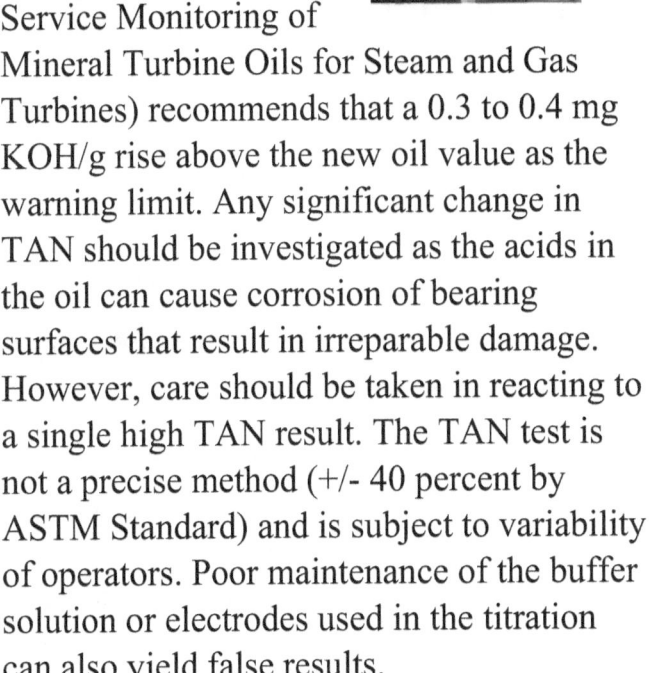

Foam Tendency and Stability (D892, Sequence I)

The presence of some foam in the reservoir is normal and not a cause for concern. Excessive foaming is generally not related to the oil, but rather to mechanical issues that cause excessive amounts of air to be introduced to the oil. Contamination and oil oxidation can also have an effect on the foaming tendency and stability. Excessive amounts of foam are a concern to the turbine operator for two reasons:

First is a safety and housekeeping issue if the foam overflows the reservoir. Second, excessive amounts of air in the oil can lead to more rapid oxidation and a phenomenon known as micro-dieseling. Micro-dieseling is caused when an air bubble in the oil is rapidly and adiabatically compressed causing extreme local temperature increases.

These large temperature increases are known to cause thermal and oxidative degradation of the oil leading to deposit formation.

ASTM D4378 offers the guideline of 450 ml of foaming tendency and 10 ml of stability in Sequence I test.

Colorimetric Analysis

Colorimetric analysis is designed to measure the insoluble materials in the turbine oil which often lead to varnish deposits. The process includes treating the lubricant sample with a specific chemical mixture designed to isolate and agglomerate insoluble by-product material, and collect this material on a filter patch. The color spectra of the collected material is then evaluated and depending on the intensity of specific colors or color ranges, a varnish potential rating may be derived. The filter patch may also be weighed as a means to determine insoluble concentration in the lubricant. Several commercial labs utilize this technique, each with their own specific method.

Currently, this is not covered by an ASTM standard, but an ASTM method is currently being developed based on this concept.

The color of every object we see is determined by a process of absorption and emission of the electromagnetic radiation (light) of its molecules. Colorimetric analysis is based on the principle that many substances react with each other and form a color which can indicate the concentration of the substance to be measured. When a substance is exposed to a beam of light of intensity $I0$ a portion of the radiation is absorbed by the substance's molecules, and a radiation of intensity I lower than $I0$ is emitted.

Contamination Measurement
Water Content – (Visual and ASTM D1744)

Turbine oils are subject to water contamination from several sources. Steam turbines can have leaking gland seals or steam joints. All turbines can become contaminated with water from atmospheric condensation in the reservoir or leaking heat exchangers. The turbine oil should be inspected daily for water. Looking at the sample, it should be clear and bright. A cloudy or hazy appearance indicates that water may be present. An on-site water test can be performed such as the hot plate crackle test where the subject oil is dropped on a heated metal surface. Bubbling and crackling indicate that water is present. In the laboratory, water is typically measured by Karl Fischer Titration (ASTM D1744) and reported as a percent or in parts per million.

ASTM D4378 identifies 1,000 ppm or 0.1 percent water as a warning limit.

However, some OEM's have defined 500 ppm as the warning limit. Keep in mind that the Karl Fisher method does not measure free water, so daily visual inspections of the turbine oil are recommended.

Metals by Inductively Coupled Plasma (ICP)

Metals concentration in a turbine oil can give early warning of wear conditions, changes in equipment operation or potential contamination issues. Keep in mind however, that the size of the metals detected by this method is limited to very small metal particles, typically less than 8 microns in size. That means catastrophic failures can occur where large pieces of wear metal are generated and not detected by this test. There is no specific limit on the amount of metals for turbine oils. The trend of metals concentration is often the most important aspect of this test.

Ultra Centrifuge Rating

The Ultra Centrifuge test detects finely dispersed or suspended particles in the oil. The subject 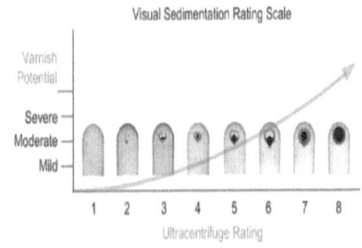 oil sample is centrifuged at 17,500rpm for 30 minutes. At the end of this period, the test tube is drained and the remaining sediment is rated.

Particle Count (ISO 4406)

Particle Counting and ISO Cleanliness ratings define the concentration of particles in the oil and relate this back to the ISO Cleanliness 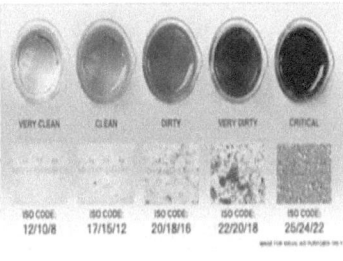 scale. The results are reported as the number of particles greater than 4 microns/6 microns/ 14 microns per ml of fluid. The ISO Cleanliness Code relates the number of particles per ml to a logarithmic scale with code number for each range. A typical result would look like 18/16/13 where 18 means there is 1,300 to 2,500 particles per ml greater than 4 microns in size, 320 to 640 greater than or equal to 6 microns, and 40 to 80 greater than 14 microns.

Particle counts are subject to a wide range of variability due to sample preparation, oil formulations, contamination of the sample container, and location and method of sampling. There are also differences in the equipment used to measure particle counts between light dispersion techniques and filter pore blockage methods. Care should be taken to ensure that the samples used for Particle Counts are representative and consistent. The particle count results are only good as a relative measure of contamination and no ASTM standard exists for this test. Ultimately, particle count does give a good indication of overall system cleanliness. OEMs do offer some guidelines for new and used oils, but in general an ISO Cleanliness code of 18/15/13 or lower is an acceptable result.

Performance Properties Test
Demulsibility (ASTM D1401)
Demulsibility is a measure of the oil's ability to separate from water. The 40 ml of the subject oil and 40 ml of distilled water

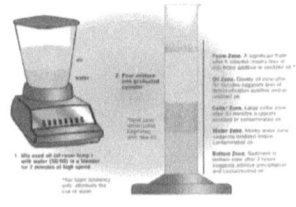

are mixed and then allowed to settle. The amount of time for full separation of the oil and water is recorded or after 30 minutes, the amounts of oil water and emulsion are recorded. ASTM does not offer a warning limit for demulsibility, but a result of 15 ml or greater of emulsion after 30 minutes is a fair warning limit. Contamination and oil age are factors that negatively affect demulsibility. Care should be taken when evaluating demulsibility as the preparation of the glassware and the quality of the water used can yield false or failing result.

Suggested Oil Analysis Schedule for Turbine System

Test	Steam	Gas	Frequency
Viscosity-ASTM D445	x	x	Monthly
TAN- ASTM D664	x	x	Monthly
RPVOT (oxidation test) - ASTM D2272	x	x	Quarterly
Water content (visual)	x	x	Daily
Water content-ASTM D1744	x	x	Monthly
Particle count	x	x	Monthly
Rust test	x	x	Only if corrosion issue
Foam	x	x	Only if foam is an issue
Demulsibility	x	x	Only if water separation is concern
Ultra centrifuge	x	x	Monthly to Quarterly
Varnish potential rating		x	Monthly to Quarterly

Troubleshooting the Oil Analysis Result for Turbine

Low Viscosity	High Viscosity	TAN	Metals	Particle Count	Water
Low viscosity oil used as make-up	Higher viscosity oil used as make-up	Increasing or high oxidation	Inaccurate sample (bottom sample)	Inaccurate sample	Atmospheric condensation
Mechanical shear in VI improved oils	Excessive oxidation	Wrong oil	Component wear		Leaking oil coolers
Contamination with solvents	Hot spots within the system	Contamination with a different fluid	Wrong oil	Filtration equipment not operating properly	Ingress of water wash

Low Viscosity	High Viscosity	TAN	Metals	Particle Count	Water
Thermal cracking from excessive heat (such as electric tank heaters)	Over extended oil drain interval	Testing variability	Sealants	Poor storage and handling procedures	Steam leaks
Bad or mis-labeled sample	Contamination		Thread compounds		Poor oil demulsibility

Low Viscosity	High Viscosity	TAN	Metals	Particle Count	Water
	Bad or miss-labeled sample		Contaminants		Oil conditioning equipment not functioning properly
			Assembly lubes		Inaccurate sample (bottom samples)

Case Study.3: Oil Condition Monitoring for Electrical Components

Test	Standard ASTM Reference
Dielectric Breakdown Voltage, KV	D877 D1816
Water Content, Maximum PPM	D1533
Power Factor, %	D924
Interfacial Tension, dynes/cm	D971 D2285
Acidity, mgKOH/gm	D974

According to the American Society for Testing Materials

1. Dielectric Breakdown:

The dielectric breakdown voltage is a measurement of electrical stress that an insulating oil can withstand without failure.

2. Power Factor:

The power factor of insulating oil equals the cosine of the phase angle between an ac voltage applied and the resulting current. Power factor indicates the dielectric loss of the insulating oil and, thus, its dielectric heating.

3. Interfacial Tension

The interfacial tension (IFT) test is employed as an indication of the sledging characteristics of power transformer insulating oil. It is a test of IFT of water against oil.

4. Water Content

Actual water percentage exist in the oil sample.

Gas Dissolved Tests (according to ASTM standard)

Gases		Allowed percentage in 11KV Transformer
Acethylene1	C2H2	From Oil Table depend on the oil type used.
Nitrogen	N2	
Carbon monoxide1	CO	
Oxygen	O2	
Methane1	CH4	
Carbon Dioxide	CO2	
Ethane1	C2H6	
Ethylene1	C2H4	
Hydrogen1	H2	

Troubleshooting Dissolved Gas Analysis (Example)

Gas	Normal	Abnormal	Problem
H2	<150 ppm	>1000 ppm	Corona, Arcing
CH4	<25 ppm	>80 ppm	Sparking
CO	<500 ppm	>1000 ppm	Severe Overloading
CO2	<10000 ppm	>150000 ppm	
C2H4	<20 ppm	>100 ppm	Severe Overheating
C2H6	<10 ppm	>35 ppm	
C2H2	<15 ppm	>70 ppm	Arcing

Oil Analysis Interval for Transformers

Rating	Type of Transformer	Period Between Testing (months)	
		Gas	Oil
>1MVA	Furnace (critical)	3	6-12
	Distribution	6	6-12
	Special	3-6	6-12
<1MVA	Any type	6-12	12

Portable Oil Analyzer Applications: All types of engines include car, trucks,

loaders, turbines, and generators.

Quick indication for:
- ✓ TAN.
- ✓ TBN.
- ✓ Oxidation degree.

Thermography Analysis

Introduction about Thermography Technique
The Technique
Infrared monitoring and analysis has the widest range of application (from high- to low-speed equipment), and it can be effective for spotting both mechanical and electrical failures. It also requires minimum skills for analysis.

What is infrared radiation?
Everything on this planet contains thermal energy and therefore has a specific temperature. This thermal energy is emitted from the surface of the material. This energy is called is infrared (IR) radiation. The amount of IR radiation emitted at a certain wavelength, from the surface of an object, is a function of the object's temperature. This is a very important concept, since it implies that one can calculate the temperature of an object by measuring the infrared radiation emitted from it.

How is infrared energy related to problem detection?

Detectors in the infrared camera convert this incoming infrared energy from the infrared spectrum to the visual spectrum so we can see the infrared energy.

What is Infrared Thermography?

Infrared Thermography is the technique for producing a visible image of invisible (to our eyes) infrared energy emitted by objects.

Thermograph spans many areas

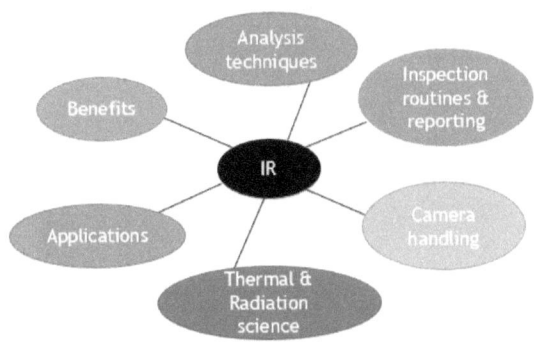

Applications of Thermography
- Condition based maintenance.
- Research & Development.
- Medical and veterinary.
- QC and process monitoring.
- Non-destructive testing.

Measurements are:
- Non-contact.
- Obtained without disturbing production.
- Applies to all type of equipment.
- Reliable data
- Quickly identifies specific location
- Apply to most all conditions

Average downtime cost:
Lost Revenues

Industry Sector	Revenue/Hour
Chemicals	$704,101
Construction & Engineering	$389,601
Electronics	$477,366
Energy	$2,817,846
Food & Beverage	$804,192
Manufacturing	$1,610,654
Metals/natural resources	$580,588
Pharmaceuticals	$1,082,252
Utilities	643,250

Various Inspection Points

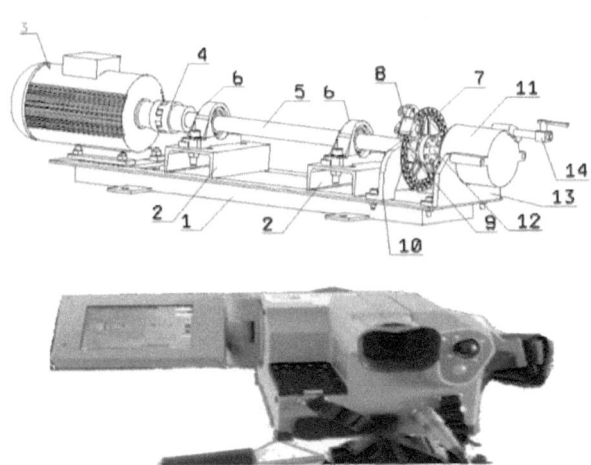

Machine Heat Sources

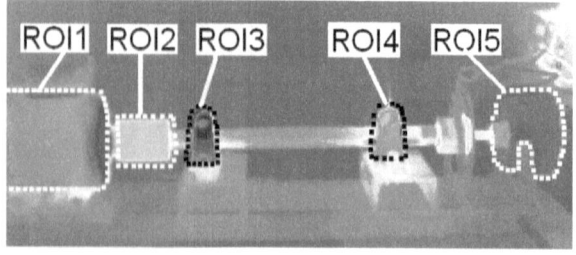

Industrial Equipment Applications

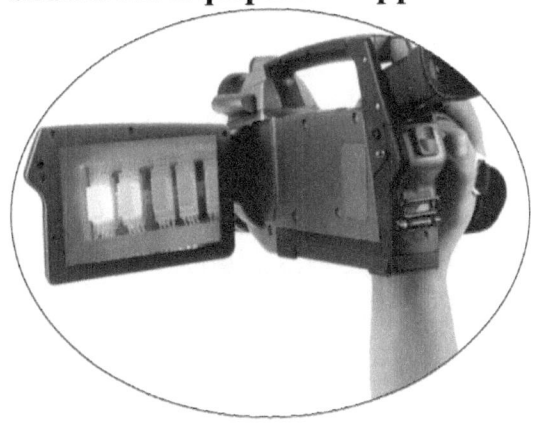

Electromechanical & Mechanical Systems

- ✓ Pumps
- ✓ Fans
- ✓ Heat Exchangers
- ✓ Gearboxes
- ✓ Bearings
- ✓ Drive belts
- ✓ Motors

Possible reasons for temperature hotspots or deviations:

- Bad alignment, overload, bent shafts…etc.

- Friction due to wear, misalignment or inadequate lubrication

Before you start up!

Tips:
- ✓ Make reference for the temp of all equipment when installing them in a new plant by taking photos and recording temperatures.
- ✓ If the above step were skipped or haven't been performed, ask the manufacturer for the temp reference of all equipment parts under normal operation.
- ✓ If the above tip is not possible, use the comparison method by comparing a pump with the one beside it, same for any other equipment fans, compressors, transformers…etc. Use the standby maintained equipment as a reference.

Bearing Failures

Bearing Temp is 108.4degC due to overload issue and this may be a cause of unbalance issue

Overall bearing temp is high due to greasing problem

Bearing temp is 81.2

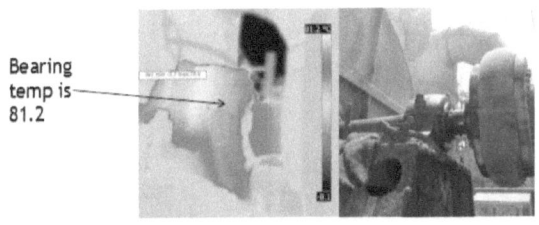

Clean the surface and recheck again

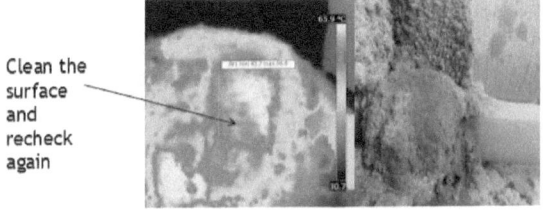

Infrared Thermography need a relativity clean service to avoid image errors

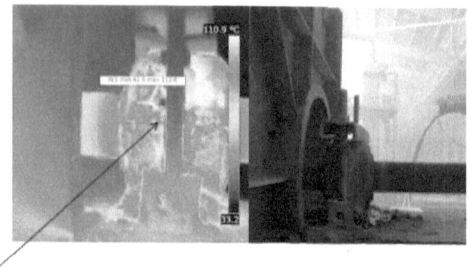

Bearing temp is 110degC due to overload issue, this may be due to unbalance issue.

Bearing in centrifugal fan shaft, possible cause of overload is unbalance or misalignment. Early detection of problem mean saving bearing life, shaft, and maintenance costs. Also unexpected plant shutdowns.

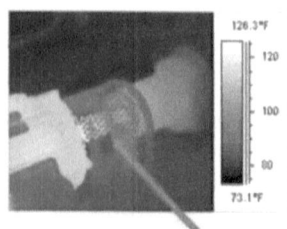

BEFORE REPAIR
SUSPECT WORN COUPLER/SPIDER

Label	Value
SP01	110.0°F

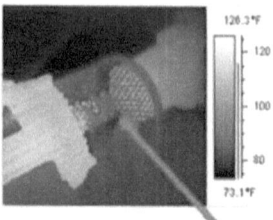

AFTER REPAIR
NOTE COOLER COUPLER DELTA-T

Label	Value
SP01	87.8°F

Motor belt is overheated, possible reasons are loose belt or over tightened

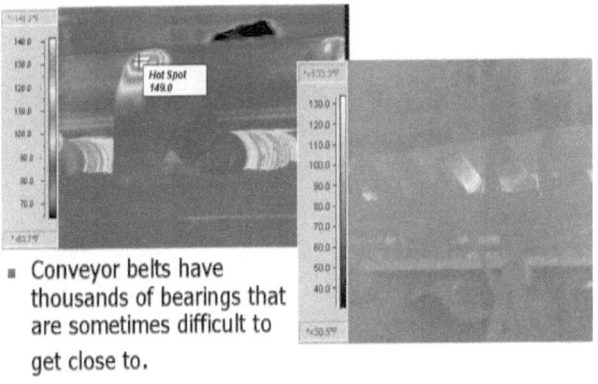

- Conveyor belts have thousands of bearings that are sometimes difficult to get close to.

Infrared thermography facilitate inspection, it's really a fast and reliable method for inspecting thousands of bearings and equipment every day.

In a belt conveyors, failure of roller bearings can cause the rubber belt to wear overtime. The consequence of failure for those small bearings can be huge at the end. A conveyors is used to convey materials. Breakdowns can delay productivity, therefore delay the product to customer result in unhappy customer or losing few customers!

Pump Bearing

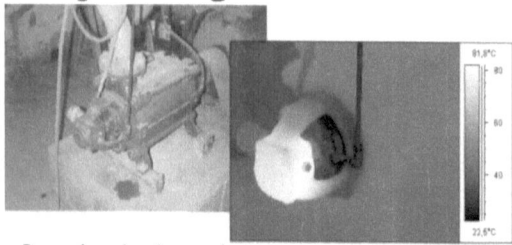

- Pump bearing in need of maintenance.

Electric Motor Bearing

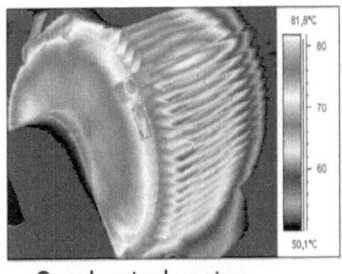

- Overheated motor bearing. Over 80 °C on bearing housing.

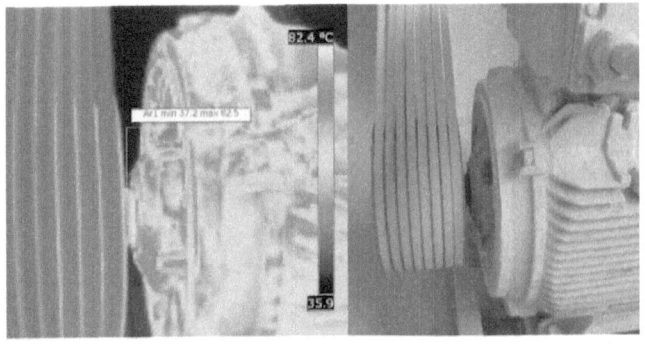

Motor bearing temperature is high due to excessive belt tension

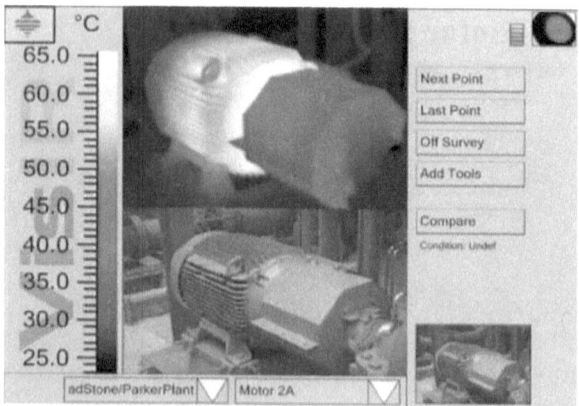

Overheated Motors

Heat Exchanger: blocked Heat exchanger

Thermography Process Installations Application

- Metal Foundry
- Manufacturing equipment
- Pipes and valves
- Plastics Industry (Molding)
- Steam systems/traps
- Refractory insulation
- Tanks and vessels
- Boilers and Reactors
- Heaters/Furnaces

Furnace Refractory

Thermal Isolation

Furnace refractory damage

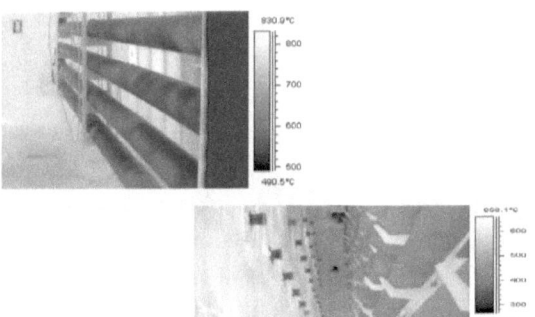

Furnace tubes and burners

Boiler

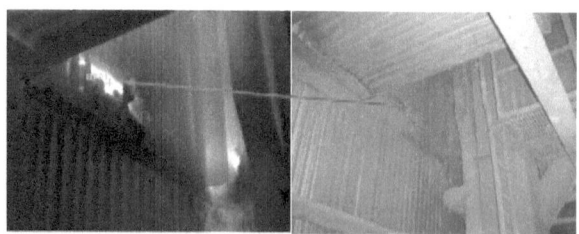

Casing leaks such as this one can be easily identified with infrared

Turbine Problems and Leakage

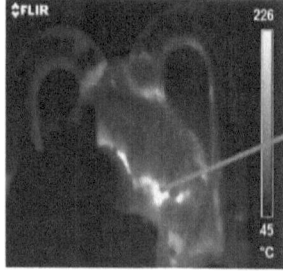

Cement rotary kiln

Piping: lack of insulation

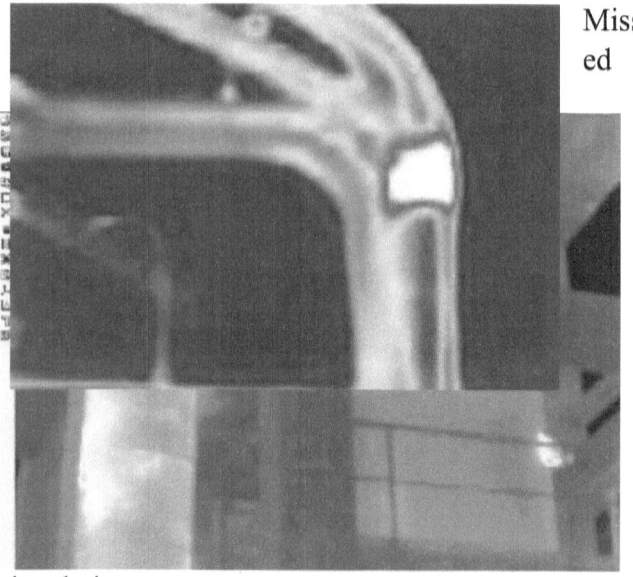

Missed insulation

Condenser Air Leakage

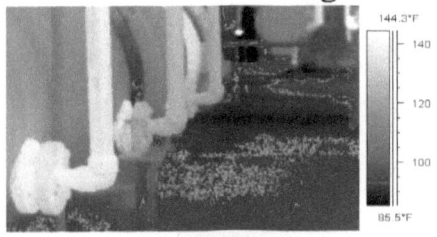

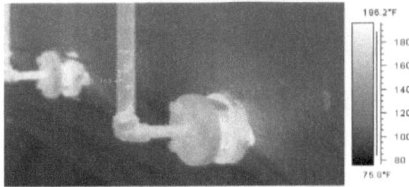

Temp different in flanges indicate leakage

Detect Valves Leakage

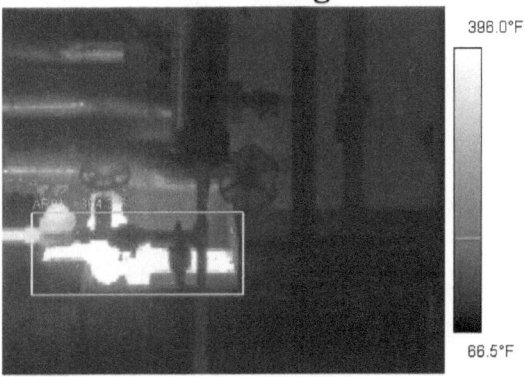

This leaking valve is over 177oC (350oF). Most leaking valves are less evident.

All boiler, turbine, stop valve, valve chest, etc., drain lines need to be checked for leak through.

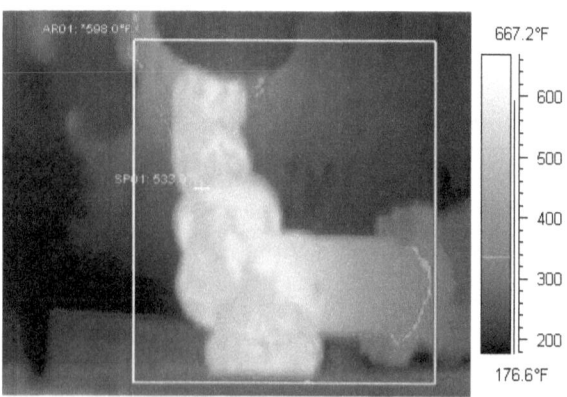

Turbine reheat steam line drain valve leaking through.

Note: Make sure a valve is totally closed before inspecting

 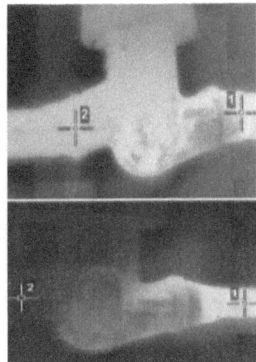

Power plant valves leakage

Steam Traps

- ✓ IR can identify leaking bypass lines and improper operation (must monitor).

Steam trap stuck open.

Steam trap working normally.

- ✓ One must know the trap cycle of operation.
- ✓ Comparison between like equipment that is operating the same often confirms problems.

Steam trap by-pass leaking.

- ✓ Use Ultrasonic Acoustics to confirm problems.

Storage tanks levels

Electrical Systems Application
- (3 phase) Power distribution
- Fuse boxes
- Cables & connections
- Relays/Switches
- Insulators
- Capacitors
- Circuit breakers
- Controllers
- Transformers
- Battery banks
- Motors
- Substations

Motor Starter Connection Problem

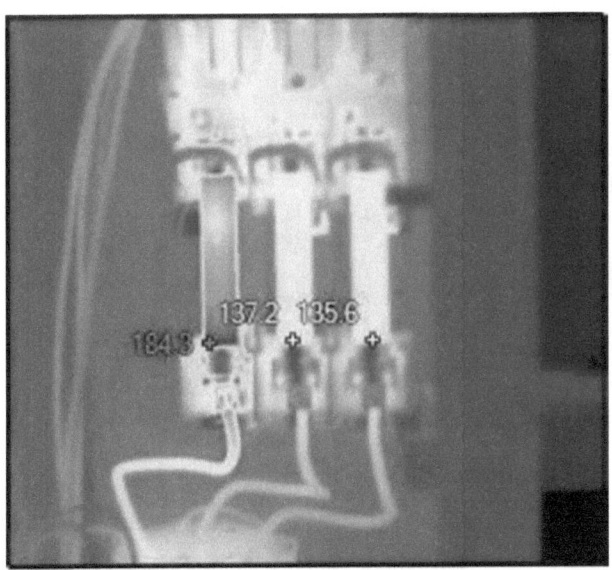

Higher temp for one starter compared to the other two

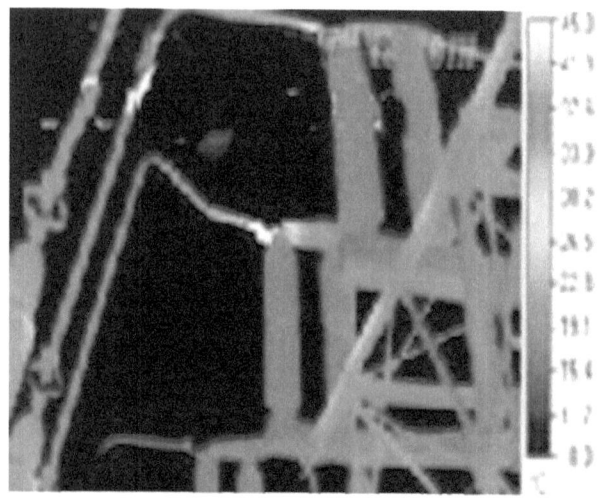

Connection failure in an electrical component

Transformer Problems

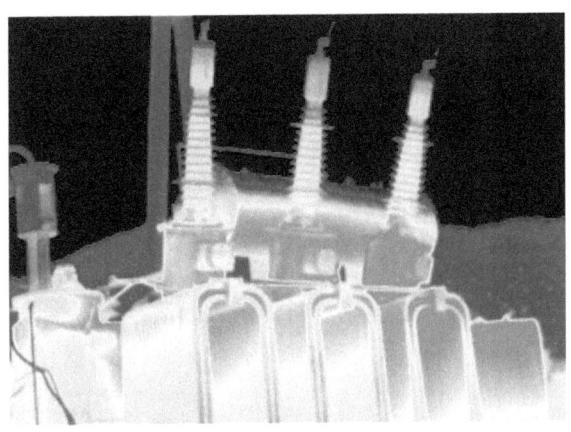

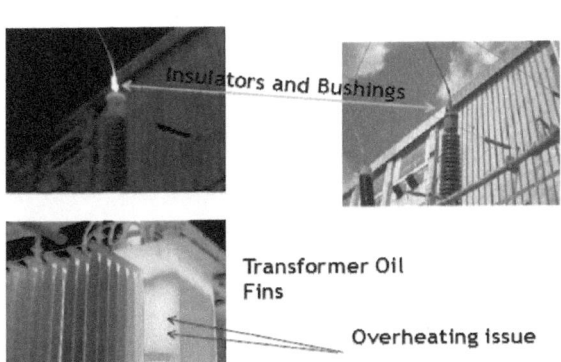

Insulators and Bushings

Transformer Oil Fins

Overheating issue

Transformer Winding problem

Thermo graphing Windings

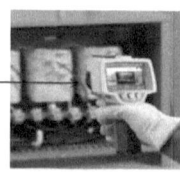
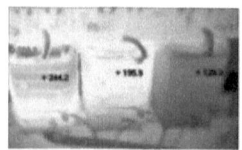
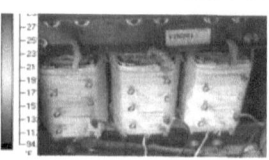

Capacitor

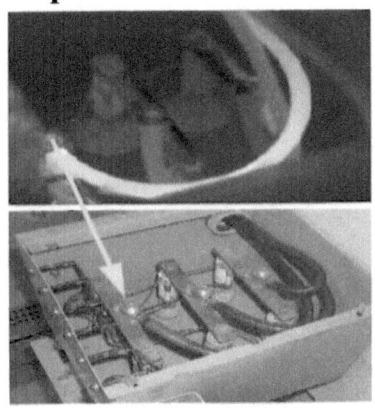

Electric Insulation Problem

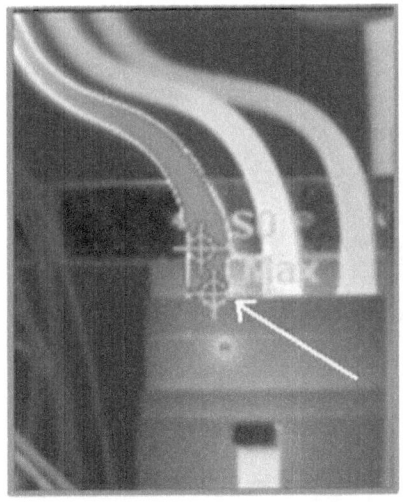

Infrared Analysis for Electric Panels

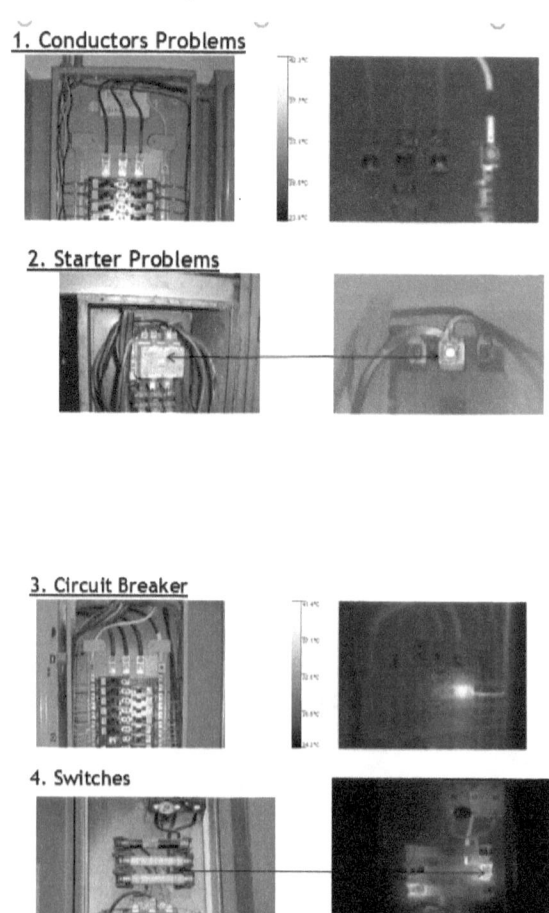

1. Conductors Problems

2. Starter Problems

3. Circuit Breaker

4. Switches

5. Terminal Connectors

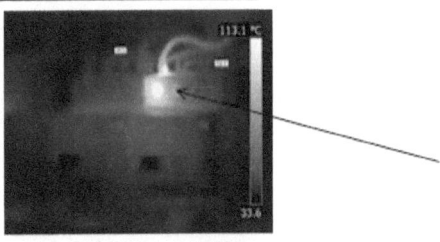

6. Switch Gears

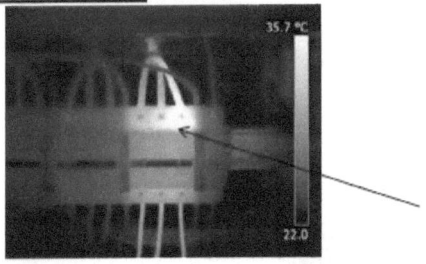

7. Breaker wires & connectors

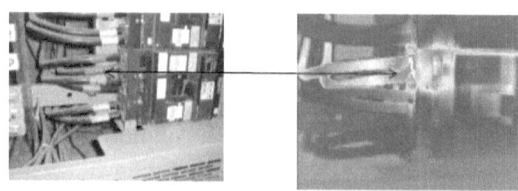

8. Conductors

9. PLC (Some panels have controllers)

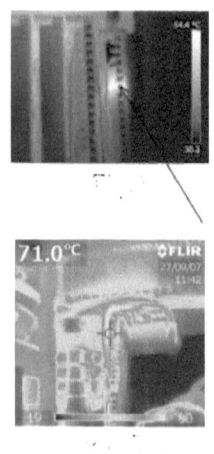

10. Fuse Boxes

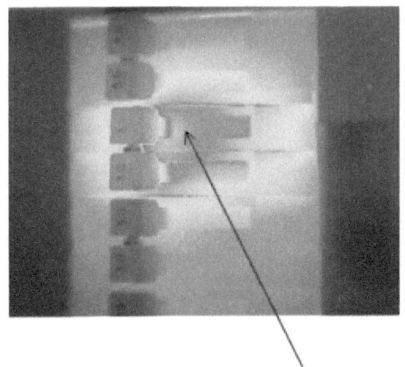

Buildings Inspection

Commonly inspected components:
- Walls
- Roofs
- Windows
- Doors
- HVAC
- Insulation
- Floor heating

Concrete Inspections

This example shows that even though the bridge deck doesn't generate heat it can still be analyzed with thermography.

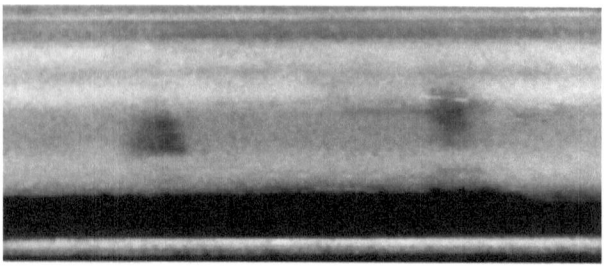

House Insulation Inspection

Missing insulation in a residential dwelling wall costs the occupants hundreds of dollars each year in heating and cooling.

Air Leakage around the Roof Latch

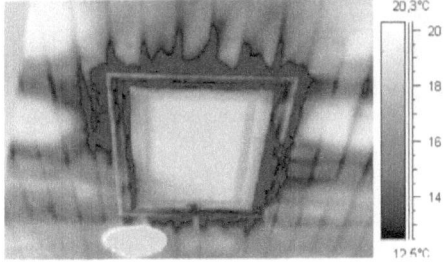

Building –Moisture in Roof

Roof (trapped water)

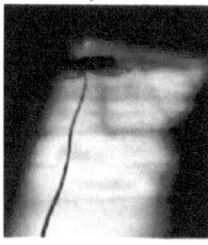

Drain Blockage at House

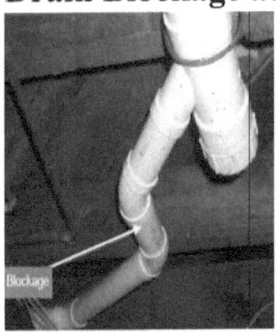

Heat-loss in Building

Aircraft Inspection

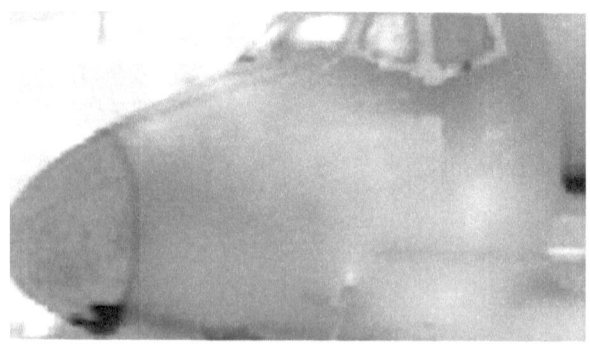

Composite aircraft materials are extremely sturdy and lightweight. These materials are vital to aircraft performance and airworthiness. However, the honeycomb structure of this material presents a potentially dangerous problem: water ingress.

Medical Applications

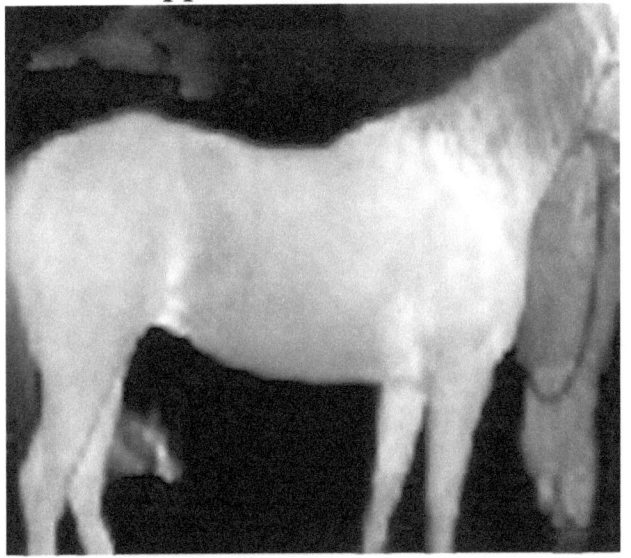

Race horse sustained an injury in a fall. The infrared image shows where the problem is, and monitored the process of the healing.

Some application for human dieses:
Breast Pathologies

It's used to screen for breast cancer, benign tumors, mastitis, and fibrocystic breast disease.

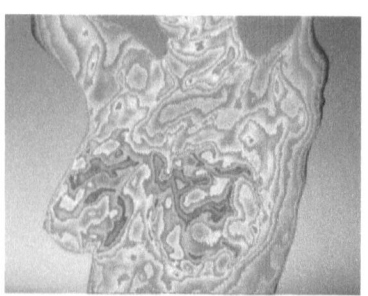
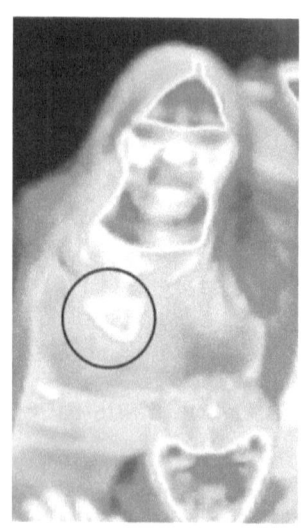

Thermography can spot breast cancer

Neuro-Musculo-Skeletal

This is where medical thermography shines the most. In this application, a thermal camera can showcase its ability to accurately diagnose patients that are experiencing problems with their back, neck, and extremities.

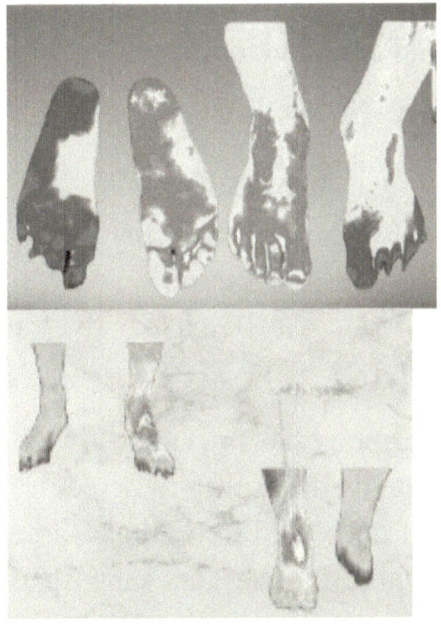

Detect muscle problems and foot complications

Vertebrae (Nerve Damage or Arthritis)
This test of function relies on the sympathetic nerve control of skin blood flow and the ability of the sympathetic system to respond and react to pathology anywhere in the body. In simple terms, this test can indicate where there are inflammatory conditions in the body and show up early signs of abnormal cell development.

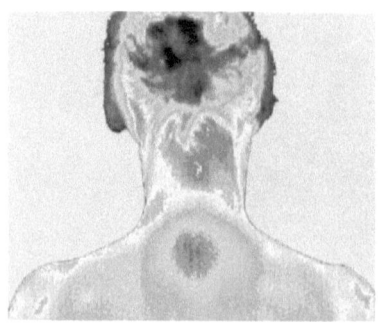

Lower Extremity Vessel Disease

Medical Thermography has provided medical professionals the ability to detect any semblance of deep vein thrombosis, as well as other circulatory disorders safely and painlessly. If you leave those disorders unchecked, it might lead to the loss of limbs, and possibly a stroke.

Extra-Cranial Vessel Disease

In similar fashion to breast pathologies, different conditions that involve the flow of blood through the vessels in the neck and head area are easily diagnosed with thermal imaging.

Fever Screening

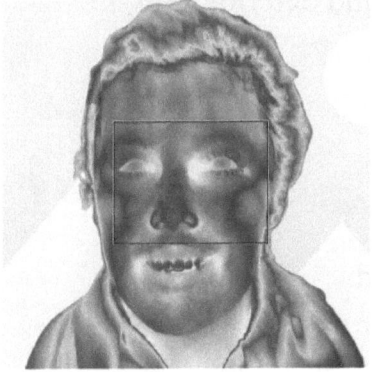

Advantages of Thermography and Limitations

- ✓ It shows a visual picture so temperatures over a large area can be compared.
- ✓ It is capable of catching moving targets in real time.
- ✓ It is able to find deteriorating, i.e., higher temperature components prior to their failure.
- ✓ It can be used to measure or observe in areas inaccessible or hazardous for other methods.
- ✓ It is a non-destructive test method.
- ✓ It can be used to find defects in shafts, pipes, and other metal or plastic parts.
- ✓ It can be used to detect objects in dark area.

Limitations
- ✓ Due to the low volume of thermal cameras, quality cameras often have a high price range (often US$6,000 or more).
- ✓ Images can be difficult to interpret accurately when based upon certain.
- ✓ Objects, specifically objects with erratic temperatures, although this problem is reduced in active thermal imaging.
- ✓ Accurate temperature measurements are conflicted by differing emissivity and reflections from other surfaces.
- ✓ Most cameras have ±2% accuracy or worse and are not as accurate as contact methods.
- ✓ Only able to directly detect surface temperatures.

Infrared Thermometer – A simple hand held device

- Rapid, precise temperature measurement.
- State of the art infrared temperature
- Measurement technology.
- Simple to use
- Reduction of unplanned downtime
- Available at economical cost.

Terminologies

- Emissivity.
- Distance to spot.
- Temp range.

Emissivity

Emissivity is a term used to describe the energy-emitting characteristics of materials. Most organic materials and painted or oxidized surfaces have an emissivity of 0.95. Inaccurate readings can result from measuring shiny or polished metal surfaces.

To compensate for this, adjust the unit's emissivity reading, or cover the surface to be measured with masking tape or flat black paint. Allow time for the tape or paint to reach the same temperature as the material. Underneath it measure the temperature of the tape or painted surface.

Distance to Spot

D: S ratio or distance to spot ratio. This indicates the size of the area measured relative to distance away from the object being measured. If D: S is 1:1 then at 1 foot the area being measured is 1 foot in diameter. If D: S is 8:1 then from 8 ft. away the area measured is 1 ft. in diameter.

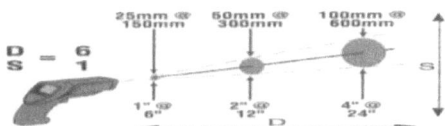

An ideal thermometer will measure from -35 to 500C, some others will measure up to 1000C.

There are two models of infrared thermometers, one is non-contact and the other is contact, the difference between them are as follow:

The non-contact is faster for measuring than the contact one. The non-contact can measure in the hazard areas where the contact one can't be used.

The only good thing in the contact one is that it can measure temperature ranges more than the non-contact one.

Emissivity Values of Common Materials

Material	Emissivity	Material	Emissivity
Aluminum, polished	0.05	Clay, fired	0.91
Aluminum, rough surface	0.07	Concrete	0.92
Aluminum, strongly oxidized	0.25	Copper, polished,	0.01
Asbestos board	0.96	Copper, commercial burnished	0.07
Asbestos fabric	0.78	Copper, oxidized	0.65
Asbestos paper	0.94	Copper, oxidized to black	0.88
Asbestos slate	0.96	Electrical tape, black plastic	0.95
Brass, dull, tarnished	0.22	Enamel	0.90
Brass, polished	0.03	Formica	0.93
Brick, common	0.85	Frozen soil	0.93

Material	Emissivity	Material	Emissivity
Brick, glazed, rough	0.85	Glass	0.92
Brick, refractory, rough	0.94	Glass, frosted	0.96
Bronze, porous, rough	0.55	Gold, polished	0.02
Bronze, polished	0.10	Ice	0.97
Carbon, purified	0.80	Iron, hot rolled	0.77
Cast iron, rough casting	0.81	Iron, oxidized	0.74
Cast iron, polished	0.21	Iron, sheet galvanized, burnished	0.23
Charcoal, powdered	0.96	Iron, sheet, galvanized, oxidized	0.28
Chromium, polished	0.10		

Material	Emissivity	Material	Emissivity
Iron, shiny, etched	0.16	Rubber	0.93
Iron, wrought, polished	0.28	Shellac, black, dull	0.91
Lacquer, black, dull	0.97	Shellac, black, shiny	0.82
Lacquer, black, shiny	0.87	Snow	0.80
Lacquer, white	0.87	Steel, galvanized	0.28
Lampblack	0.96	Steel, oxidized strongly	0.88
Lead, gray	0.28	Steel, rolled freshly	0.24
Lead, oxidized	0.63	Steel, rough surface	0.96
Lead, red, powdered	0.93	Steel, rusty red	0.69
Lead, shiny	0.08	Steel, sheet, nickelplated	0.11

Material	Emissivity	Material	Emissivity
Mercury, pure	0.10	Steel, sheet, rolled	0.56
Nickel, on cast iron	0.05	Tar paper	0.92
Nickel, pure polished	0.05	Tin, burnished	0.05
Paint, silver finish	0.31	Tungsten	0.05
Paint, oil, average	0.94	Water	0.98
Paper, black, shiny	0.90	Zinc, sheet	0.20
Paper, black, dull	0.94	Porcelain, glazed	0.92
Paper, white	0.90	Quartz	0.93
Platinum, pure, polished	0.08		

How to Choose the Right Camera

Important Specifications:

- Temp Measuring Ranges: -30 to 1500C
- Thermal Sensitivity: 100mkv - < 50mk
- Resolution: from 160x120 to 320x240
- Accuracy: 2% or less
- Emissivity: adjustable from 0.01 to 1

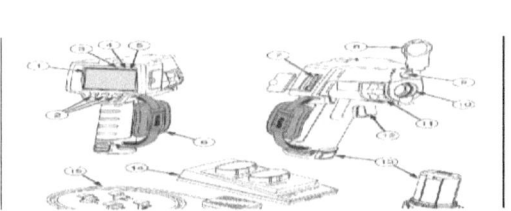

1-LCD display	7-SD memory card	13-Removable lithium battery
2-Function soft keys	8-Lense cover	14-Two-Bay charging Base
3-Speaker	9-Visual light	15-AC adapter/Power Supply
4-Microphone	10-Infrared lens	
5-Autobacklight sensor	11-Focus control ring	
6-Hand strap	12-Image capture trigger	

The menus, coupled with the three soft keys (□, □, and □), provide access for thermal image display,

saving and viewing stored images, and setting features:
• Backlight
• Date/Time
• Emissivity
• File Format
• High Temperature Alarm (Ti32, Ti29, Ti27) or Dew-point Alarm (TiR32,
TiR29, TiR27)
• Hot Spot and Cold Spot and Center Point on the image
• IR-FusionR Mode
• Language
• Lens Selection
• Level/Span
• Palette
• Reflected Background Temperature Compensation
• Temperature Scale
• Transmission Correction

Reporting and Software

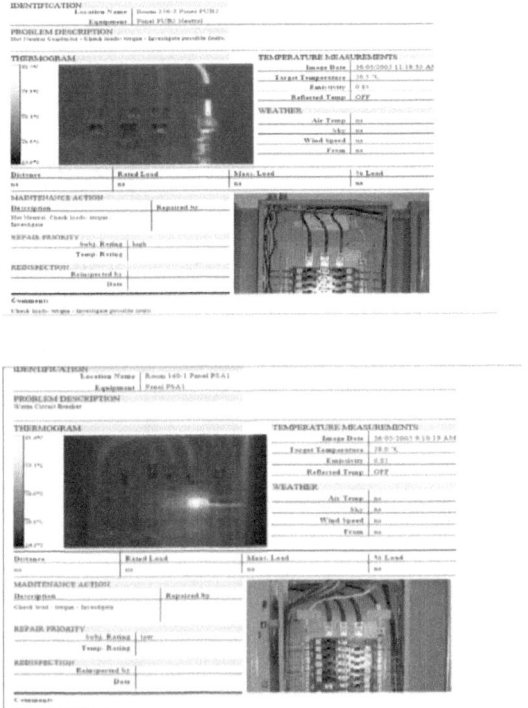

Consideration when photo graphing images:

- As greater as you go close to the object being inspected when using an infrared camera; while taking safety concerns, the more likelihood of an inaccurate reading.

- Long-distance lenses are used for inspections that preclude close proximity, such as electrical poles. A wide-angle lens is used to inspect large objects.
- A variety of weather conditions affect infrared thermography readings; for example, high winds during a roof inspection can obscure the imaging results. Other weather conditions that must be taken into effect include summer sunlight and rain.
- The inspector must adapt to weather conditions to ensure accurate readings, including avoiding testing during times of rain or waiting for gusts of wind to subside. Avoid testing during times of intense sunlight, such as the middle of the day. Reschedule testing if adjustments to the conditions are not possible.
- If the surface you are measuring is silver, white colored or reflected surface, please refer to emissivity

rules; you can use a darker sheet of paper and place it on the objective.

How Thermography and Lean Overlaps

How thermograph reduce equipment losses & wastes?

- Reduce Time of Maintenance.
- Detect early failure & prevent CM downtimes.
- Detect hidden failures & improve equipment reliability.
- Reduce the resources required to do the routine & preventive maintenance.

Simply, lean means creating more value for customers with fewer resources.

Lean is all about minimizing losses. Lean for equipment maintenance means minimum downtime with maximum reliability.

Ultrasound Analysis

Introduction to Ultrasound Technique
What is ultrasonic?

Ultrasound is cyclic sound pressure with a frequency greater than the upper limit of human hearing, excess of 20,000 cycles (hertz) per second (20KHZ).

So by definition, ultrasound is totally undetectable by human ears unless aided by instruments capable of translating ultrasound to audible sound. In the marketplace, these instruments are commonly known as ultrasonic detectors and have been used for various maintenance related functions for over 25 years.

Ultrasonic is a predictive maintenance technique and one of the non-destructive testing tools that used in the field of industry to detect early & hidden equipment failures.

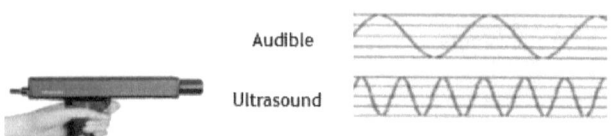

What is the different between Ultrasonic & Vibration?

Vibration is a low frequency method that can detect bearing failures and the reason of this failure.

Ultrasonic is a high frequency vibration method (ultrasonic vibration) that can detect the degrees of bearing failures & wears, it can also detect the lubrication problems of the bearing.

One of the most advantages of using ultrasonic over vibration, is that ultrasonic can reveal the lubrication problems and provide a very early warning of bearing faults.

The very early detection of bearing failure using ultrasound can save a lot of money and equipment life; preventing unexpected plant stop and loss of productivity.

Ultrasound should be a part of any predictive maintenance planning program.

Overview on the Instrument

Lightweight and portable, ultrasonic translators are often used to inspect a wide variety of equipment. Some helpful accessories are supplied with the instrument too.

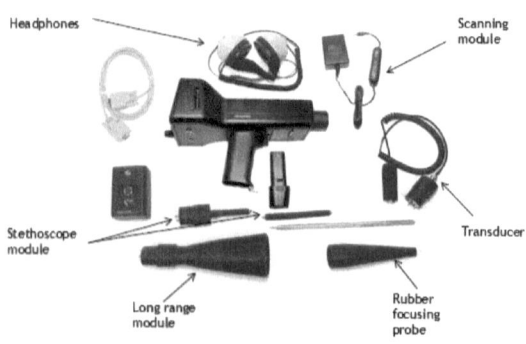

Typical Applications
- PRESSURE/VACUUM LEAKS (TURBULENCE)
- COMPRESSED AIR
- OXYGEN
- HYDROGEN ETC.
- HEAT EXCHANGERS
- BOILERS
- CONDENSERS
- TANKS
- PIPES
- VALVES
- STEAM TRAPS
- MECHANICAL INSPECTION
- BEARINGS
- LACK OF LUBRICATION/FAILURE
- PUMPS
- MOTORS
- GEARS/GEAR BOXES
- FANS
- COMPRESSORS
- CONVEYERS

- AUTOMOTIVE
- RAIL ROADS
- MARINE
- AVIATION
- ELECTRIC EQUIPMENT
- (Arcing/tracking/corona)
- SWITCHGEAR
- TRANSFORMERS
- INSULATORS
- POTHEADS
- JUNCTION BOXES
- CIRCUIT BREAKERS

Leak Detection
Reasons For
Using Ultrasound
to detect leaks:
1. Economics
2. Environmental
3. Safety

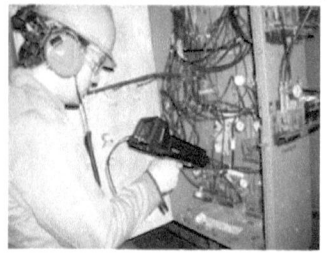
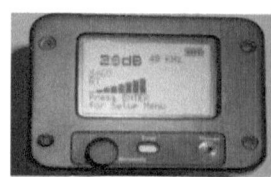

Locate the leak
Measure the Leak
Calculate costs
Calculate Greenhouse Gas emission reduction

Ultrasonic Valves Leak Detection

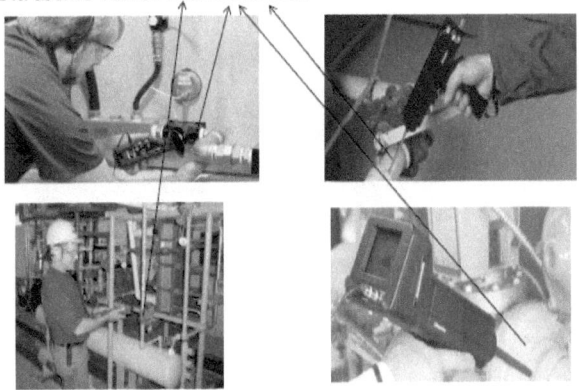

Good Valve – Bad Valve

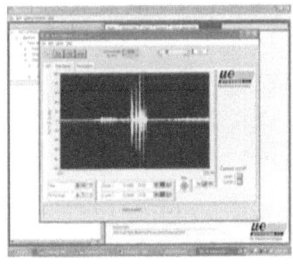
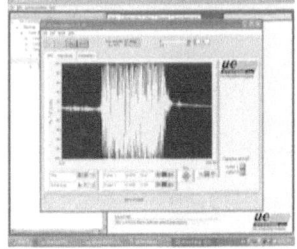

Detect Boiler leakage

Detect Heat Exchanger Leakage

Detect Steam Trap Leakage

Detect Steam Traps Leakage

Detect Condenser Leakage

Detect Tanks Leakage

Detect Compressor Leakage and Air Valves

Detect Compressed Air Leakage in an Instrument

Detect Pipes Leakage

Air Leakage

Bearings problems can be detected in any type of equipment using ultrasound

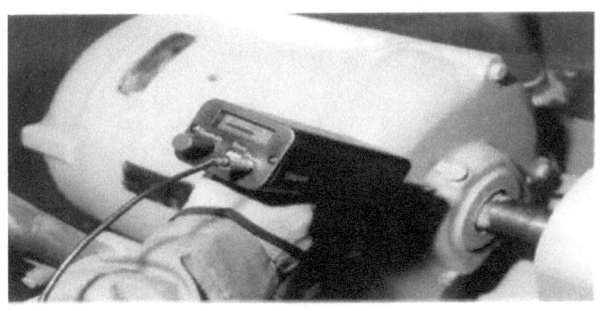

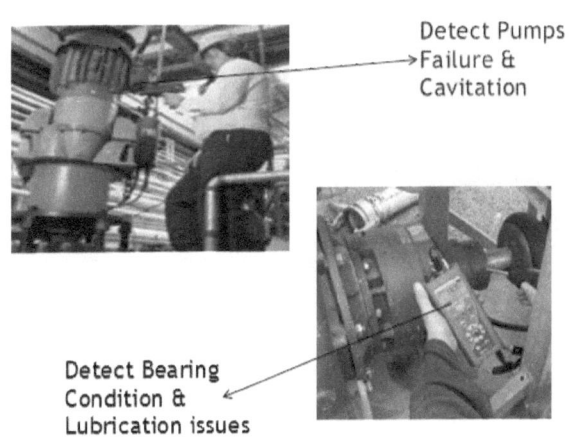

Detect Pumps Failure & Cavitation

Detect Bearing Condition & Lubrication issues

Electric Inspection

CORONA TRACKING ARCING

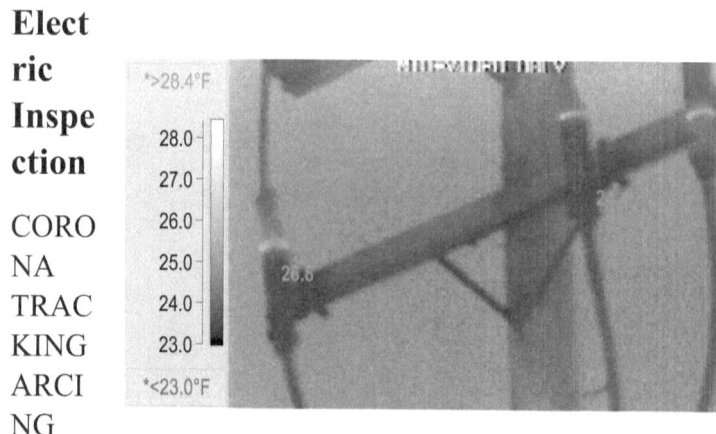

Ultrasound is Good for MEDIUM and HIGH Voltage

Circuit Breakers

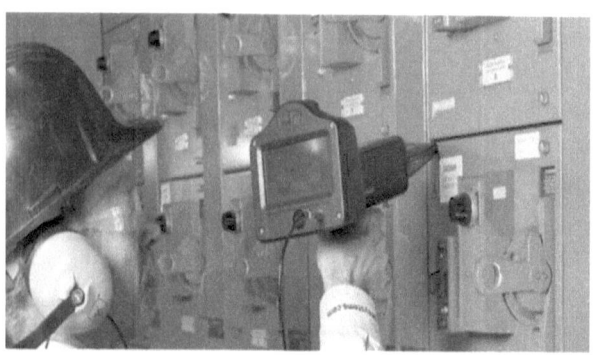

Transformer Issues: bushing, winding, oil, cooling system, condenser.

**Electric Discharges:
Arcing
Corona
Tracking**

Detect Gearbox Broken Teeth

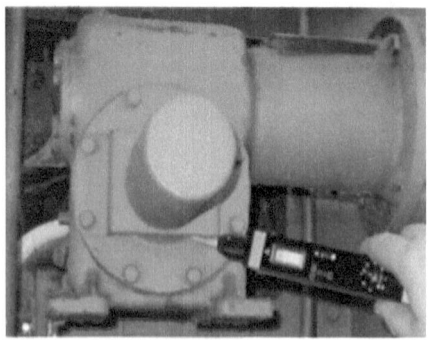

Advantages of Ultrasound
1. Ultrasound emissions are directional.
2. Ultrasound tends to be highly localized.
3. Ultrasound provides early warning of impending mechanical failure
4. The instruments can be used in loud, noisy environments
5. They support and enhance other PDM technologies or can stand on their own in a maintenance program
6. Test hazard equipment from long distances.
7. Discover early failures without stopping the equipment

Disadvantages
1. Surface to be tested must be ground smooth and clean
2. Skilled and trained operator is required.
3. Quite expensive method.

How the ultrasound instrument works?
If we want to listen to ultrasound, we need an instrument capable of translating high frequencies into a range we can hear

(normally 200-5000 hertz is a comfortable listening range). That is the function of an ultrasound detector. If we want to listen ONLY to ultrasound we need a detector with certain filters to eliminate audible or "parasite" noises. If we want to measure the energy of the ultrasound then the detector should have digital measurement capabilities.

This equipment can usually store the measurements to an onboard memory chip and transmit the data to PC software.

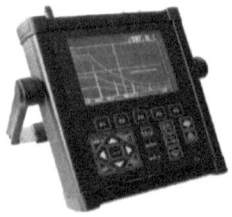 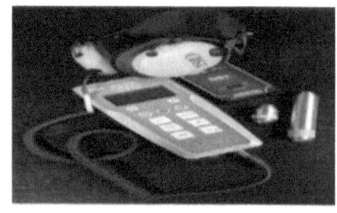

Measurement of the Signal
Ultrasonic or acoustic vibration is energy created by the friction between moving components (bearings, couplings, gear mesh, etc…). This energy is really an AC voltage or current that is at best, highly unstable and erratic. To provide useful data

for acoustic vibration monitoring this energy must be made linear for repeatability purposes. A quality ultrasonic detector uses True RMS conversion techniques to accomplish this. RMS means "Root Mean Squared." It's a way of measuring an AC voltage by means of taking the root of mean squared samples. Basically, True RMS measurement is a technique that provides consistent theoretically valid measurements of electrical signals derived from mechanical phenomena such as strain, stress, vibration, shock, expansion, bearing noise, and acoustic vibration.

The electrical signals produced by these mechanical actions are often noisy, non-periodic, and non-sinusoidal, superimposed on DC levels, and require True RMS for, valid, accurate, and repeatable measurements

Case Study:
Electric Emissions (Case Study Switch Gear inside Cabinet)

Ultrasound inspection works on all voltages, low, medium and high to detect arcing, tracking and corona in both enclosed and open access equipment. Arcing, tracking and corona ionize the air molecules around them, which produces ultrasound. With the advantage of digital sound recording and spectral analysis, inspectors can analyze sound samples to determine the type and severity of an electric emission. Below are some examples of corona, tracking and arcing. As you will note in the FFT screen as the condition becomes more severe, there are fewer harmonics of 60 cycles. If this were in Europe, we would see the same with harmonics of 50 cycles. The first image is Corona (Figure 1) followed by Tracking.

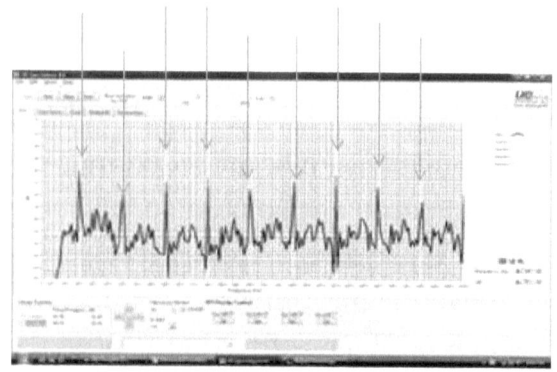

Corona

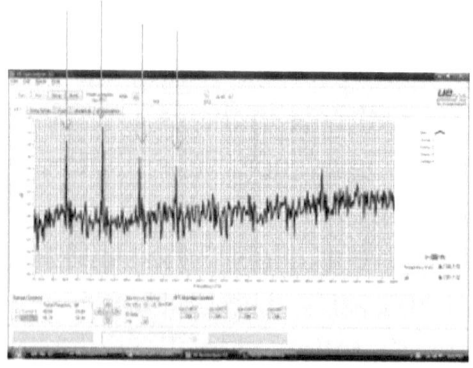

Tracking

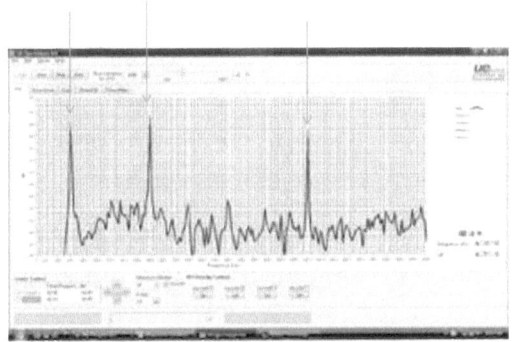

Tracking

The following demonstrate the effectiveness of ultrasound when used with infrared. An inspector who utilizes both ultrasound and infrared technologies was inspecting switchgear. Some of the doors could not be opened. There were no IR ports on the closed cabinets and therefore this switchgear could not be tested with infrared. By scanning the door seams and air vents with the ultrasound instrument, the inspector heard a very distinctive arcing sound. He recorded the sound and after the cabinets were opened he took visual and infrared images. Below are the results.

Conclusion

In extreme closed cabinet, Ultrasound beats Thermal Analysis, especially if it's not easy to open the device.

Tips

What is Corona?
Corona refers to the faint glow surrounding an electrical conductor of 3500 volts or greater as a result of the ionization of air as the nitrogen in the air breaks down. When corona occurs, it creates ozone (detrimental to the human lungs, eyes, etc.), ultraviolet light, nitric acid, electromagnetic emissions and sound.

Ozone is a strong, odorous gas that deteriorates rubber-based insulation. If moisture or high humidity conditions exist, nitric acids can also be formed that attack copper and other metals. The electromagnetic emission can be heard as interference on AM radios and the corona sound can be heard by the human ear and ultrasonic scanning devices.

Arcing

An electric arc, or arc discharge, is an electrical breakdown of a gas that produces a prolonged electrical discharge. Development of corona present Arcing.

Tracking

Tracking is the formation of partially conductive, typically carbonized, pathways on the surface of insulating materials by electrical breakdown. Development of Arcing can present tracking.

Ultrasonic Condition-Based Lubrication
Ultrasound innovation is ideally appropriate for condition-based oil techniques. With ultrasonic review instruments a program can be set up that will educate monitors which bearing should be greased up and help oil professionals realize precisely how much oil to apply.

To see how these instruments can function successfully in the loud environments of a run of the mill plant, one must comprehend the innovation of ultrasound, how ultrasound is produced by direction, and how ultrasound observing instruments can help keep up ideal oil levels in bearing.

The innovation depends on the detecting of high-recurrence sounds. Ultrasound is considered to begin at 20,000 cycles for each second, or 20 kilohertz (kHz). This is viewed as the high-recurrence edge at which human hearing stops. Most ultrasonic instruments utilized to screen hardware will detect from 20 kHz up to 100 kHz. The scope of human hearing spreads frequencies of from 20 cycles for every second (20 Hz) up to 20 kHz. The normal human will frequently hear up to 16.5 kHz and no more.

These recurrence correlations are critical to note on the grounds that there are contrasts in the manner low-recurrence and high-recurrence sounds travel, which assist us with understanding why ultrasound can be successfully established in bearing checking and oil programs.

Oiling Procedures

It is basic to think about two components of expected failure: absence of oil and over grease. Ordinary bearing burdens cause a versatile disfigurement of the components in the contact region giving a smooth circular circulation. Yet, bearing surfaces are not entirely smooth. Consequently, the genuine pressure appropriation in the contact zone will be influenced by an arbitrary surface harshness. Within the sight of a grease film on a heading surface, there is a hosing impact on the pressure conveyance, and the acoustic vitality delivered will be low. Should oil be diminished to a point where the pressure conveyance is not, at this point present, the typical unpleasant spots will connect with the face surfaces and increment the acoustic vitality. These ordinary infinitesimal distortions will start to deliver wear and the conceivable outcomes of little gaps may create which adds to the "pre-failure" condition. Accordingly, beside typical wear, the weakness or administration life of an orientation is firmly affected by the relative film thickness gave by a fitting ointment.

Staying away from Over Lubrication
At the point when an excessive amount of ointment is placed into the bearing lodging, pressure constructs up and can prompt an expansion of warmth, which can make pressure and disfigurement of the bearing. Or then again it can break or "pop" the bearing seal permitting grease to spill out into undesirable zones, (for example, an engine winding), or permit foreign

substances to enter the raceway. All of which can prompt bearing failure.

The suitable measure of oil is significant. On the off chance that a bearing is over greased up the bearing can be pushed unnecessarily by the ointment causing extra wear of the bearing. Then again, if there isn't sufficient grease, the bearing will rub on the strong surface, again causing erosion and wear on the orientation. Either case is hindering to the life of the bearing. Utilizing airborne/structure borne ultrasound removes the speculation from grease.

Ultrasound Monitoring
Ultrasound instruments identify changes identified with contact. An appropriately greased up bearing will have next to no grating. The ointment levels out any pressure the bearing experiences as it moves around the raceway in this manner decreasing the potential for ruinous contact. As the bearing moves, it delivers a conspicuous "surging" sound much the same as the sound of air spilling out of a tire. This

surging sound is alluded to as "background noise." incorporates all sounds, both low and high frequencies. The high-recurrence waves created by this background noise more restricted than those of the lower frequencies. Utilizing a ultrasonic interpreter, these signs can be identified with practically zero impedance from other mechanical commotions created by different parts, for example, a pole or another bearing close by. As the oil level in a heading falls or weakens, the potential for contact increments. There will be a relating ascend in the ultrasound adequacy level that can be noted and heard. The technique to decide when to grease up.

Furthermore, when to quit applying grease with ultrasound instruments is as straightforward as: setting a gauge, setting assessment timetables and checking as you grease up.

Setting up the greasing level:
A benchmark for an orientation reflects in decibels the level at which it is working

under typical conditions with no detectable faults and with sufficient grease.

There are three techniques for setting a pattern

1. Examination: when there is more than one orientation of a similar sort, load what's more, rpm, various orientation can be contrasted one with the other. Each bearing is examined at a similar test point and edge. The decibel levels furthermore, stable quality are looked at. On the off chance that there are no considerable contrasts, (under 8dB) a pattern dB level is set for each bearing. This is normally performed with a compact ultrasonic interpreter.

2. Set while greasing up. While oil is being applied, tune in until the sound level drops down and starts to rise. By then no more oil is included and the dB esteem is utilized as the pattern.

3. Historical: bearing dB levels are gotten from an underlying review. After thirty days the bearing dB levels are taken and looked

at. On the off chance that there is close to nothing (under 8dB) to no adjustment in dB than the benchmark levels are set and will be utilized for examination for resulting assessments.

Setting Inspection Schedules

This should be based on the equipment criticality, environment, type of industry, failure consequence, failure occurrence and the availability of standby. Typically one month is good. But for baring that have had significant levels and have been along these lines greased up, it may be important to test all the more habitually to take note of any potential changes.

Accessibility Problems

There might be circumstances in which it might be hard to access a few bearing. For instance, there might be a perplexing machine where a bearing is inserted in a territory where just a lube tube is reached out external the packaging. On the off chance that the lube tube is a conductive metal, for example, copper, the bearing can even now be tried and a grease activity level set. On the off chance that the fitting is of a non-sound conductive material, for example, plastic, a different conductive metallic wave guide can be introduced so the bearing can be observed. The wave guide can be confined from structure borne clamor of the machine (the mounting point) through elastic disconnection material. Should it not be conceivable to put a wave manage, there is an elective arrangement. A transducer can be forever mounted on the bearing lodging and a link race to an opening. The link can be appended to a specific connector that can be "connected" to the ultrasonic sensor, as demonstrated as follows.

Auto greaser and ultrasound sensor

Some systems prefer to use auto greasers for bearing. In this case installing an ultrasound sensor is a must to adjust the grease flow. The grease will be on and off based on the ultrasound sensor orders.

Conclusion

Ultrasound innovation is perfectly appropriate for viable condition-based grease programs. The short wave nature of the sign diminishes obstruction from contending commotions and permits reviewers to precisely screen bearing condition. By setting up a caution level of 8 dB over a given benchmark, investigators will know when and when not to grease up. Over oil can be dodged by applying just enough grease to accomplish standard levels or tune in to a drop in the sound level should no dB reference be accessible.

Faults Detection Using Ultrasound

Please refer to this link for a complete set of sound tracks of the most common faults detects by ultrasound.

https://drive.google.com/file/d/11G9G9QKR6Eu789At2TAGVOTkK4XHfwio/view?usp=sharing

References:

Scheffer, C. and Girdhar, P. 2004. Practical Machinery Vibration Analysis and Predictive Maintenance: Newnes; 1st Edition Predictive.

MOBIUS Institute (ilearn).

American Vibration Institute.

Scheffer, C. and Girdhar, P. 2004. Practical Machinery Vibration Analysis and Predictive Maintenance: Newnes; 1st Edition Predictive.

Oil Analysis Basics: Second Edition. Noria Corporation.

Noria Oil Sampling Procedures. Noria Corporation.

https://thermogears.com/applications-of-thermal-imaging-in-medical-science/

B.B.LahiriS.BagavathiappanT.Jayakumar John Philip. Medical applications of infrared thermography: A review. Infrared Physics &

Technology. Elsevier. Volume 55, Issue 4, July 2012, Pages 221-235.

Scheffer, C. and Girdhar, P. 2004. Practical Machinery Vibration Analysis and Predictive Maintenance: Newnes; 1st Edition Predictive.

https://www.novuslight.com/infratec-thermographic-cameras-for-fever-detection_N10340.html

https://www.thermographyofmontana.com/thermography-sports-medicine/

UE Systems Lubrication E-Book.

The Path to Lubrication Excellence – UE Systems.

Murphy, T. & Reinstra, A. 2010. Hear More A Guide to Using Ultrasound for Leak Detection and Condition Monitoring. Reliabilityweb.com; 1st Edition.

About the Author

Mohammed Hamed Ahmed Soliman is an industrial engineer, consultant, university lecturer, operational excellence leader, and author. He works as a lecturer at the American University in Cairo and as a consultant for several international industrial organizations.

Soliman earned a bachelor of science in Engineering and a master's degree in Quality Management. He earned post-graduate degrees in Industrial Engineering and Engineering Management. He holds numerous certificates in management, industry, quality, and cost engineering.

For most of his career, Soliman worked as a regular employee for various industrial sectors. This included crystal-glass making, fertilizers, and chemicals. He did this while educating people about the culture of continuous improvement.

Soliman has lectured at Princess Noura University and trained the maintenance team in Vale Oman Pelletizing Company. He has been lecturing at The American University in Cairo for 6 year and has designed and delivered 40 leadership and technical skills enhancement training modules.

Soliman is a member at the Institute of Industrial and Systems Engineers and a member with the Society for Engineering and Management Systems. He has published several articles in peer reviewed academic journals and magazines. His writings on lean manufacturing, leadership, productivity, and business appear in Industrial Engineers, Lean Thinking, and Industrial Management. Soliman's blog is www.personal-lean.org.

www.ingramcontent.com/pod-product-compliance
Lightning Source LLC
Chambersburg PA
CBHW020634220526
45464CB00001B/149